A Thousand Nights with Strangers

Confessions of a Left Coast Rideshare Driver

Burt Hays

ISBN: 978-0-9834902-3-4

Primitive Road Publishing
Flagstaff, Arizona

Acknowledgements

I'm grateful to the following individuals who contributed to the creation of this book.

Jamie Siegel was the person who first encouraged me to write these stories down and helped with some Santa Cruz details along the way.

Patti Lynn Wilson told me she could envision a book when I shared a few early versions of these stories with her.

Dr. Chris Gunn listened to some of these tales one night as we each read parts of our individual memoirs, and convinced me the book might have some commercial potential.

Mary Sojourner and Dennis Damore each read excerpts and shared their insights with me.

Burt Gershater offered some suggestions on how to organize the chapters.

Mark Statman and David Scott Moyer, two friends in Oaxaca and both outstanding writers themselves, encouraged me in 2024 to get back to work and finish writing the book.

Dr. Paty Garcia, a fellow veteran and great friend, assisted me in learning to navigate Mexico, believed in me, was available 24/7, and helped me hold it together as America was falling apart.

Both E. Shantie Ouellette and Simran Khalsa lent their editing skills, finding the typos, errors and inconsistencies, as well as improving the tone and flow of the stories.

Kirin Khalsa created the map of the Santa Cruz area that provides a wonderful cohesion to these stories, and helps make sense of the many locations in and around Santa Cruz.

Maps of the San Lorenzo Valley and the Big Picture were based on maps from Google.

Both Simran and Kirin were of great help in the development of the cover graphics.

All song lyrics were written by the author.

About the Author photo by Sean Openshaw Photography

As always, thanks to the team at Primitive Road Publishing.

Table of Contents

Map of the Santa Cruz Area

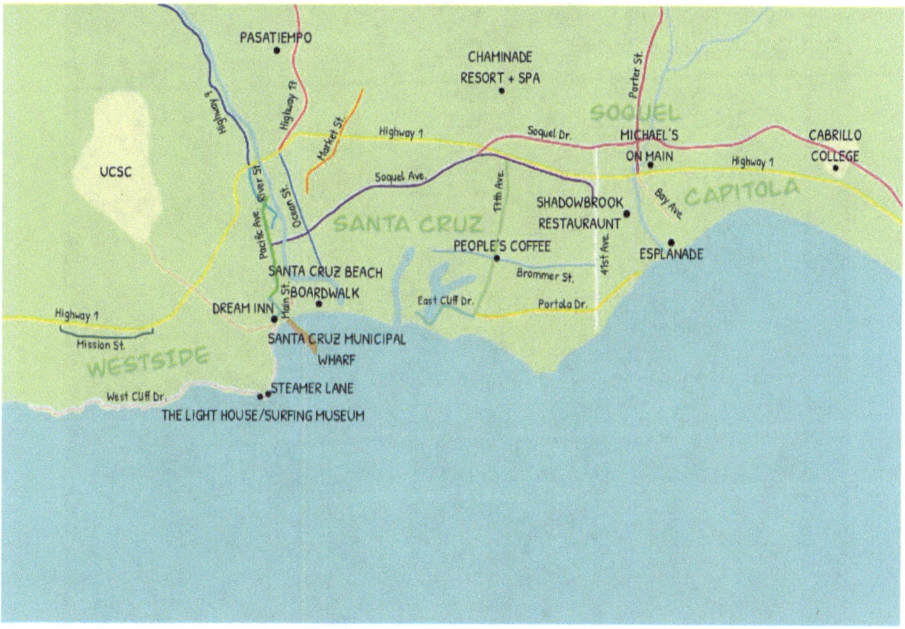

Map of the San Lorenzo Valley

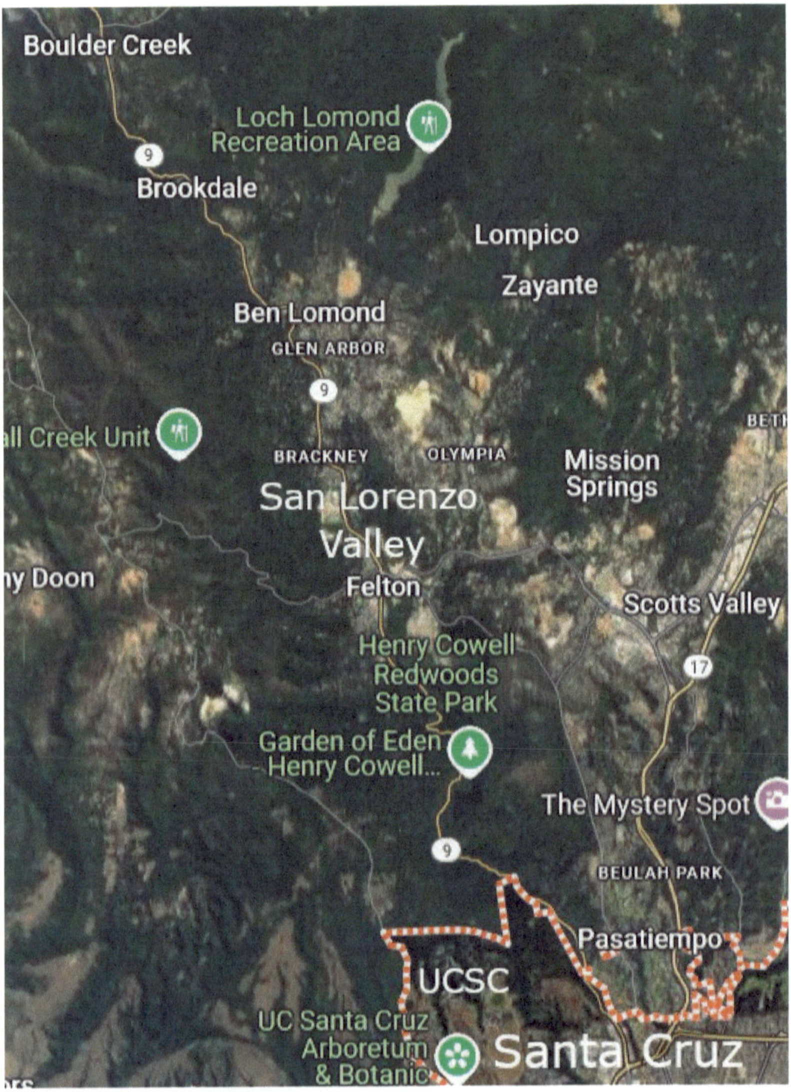

The Big Picture Map

Prologue

Truth or Dare, Part 1

I was driving for Lyft on Saturday night
That crazy boardwalk beach town scene
I was feeling like Harry Chapin
Maybe you know what I mean

From the song *Favorite Souvenir*

It was March of 2017. Trump had taken office and named Uber founder and CEO Travis Kalanick to his business advisory council. This triggered a response from lots of pissed off people, many of them students, encouraging rideshare customers to delete the Uber app. In Santa Cruz (the leftmost city on the Left Coast), Uber business took a nosedive. So, on that particular night, most of my customers were on the Lyft app. She was one of them.

I picked her up on a Saturday night at the Dream Inn, the big hotel on the beach to the west of the Santa Cruz Municipal Wharf. She sat next to me in the front passenger seat. Let's call her Cassandra.

She wore her long dark hair pulled back, tied with a black barrette with some small, turquoise stones. Her features might have been Hispanic; then again, maybe not.

"Nice hat," Cassandra said.

"Stetson," I replied with a smile.

"Nice shirt. The parrots go great with the palm trees."

“Thank you.” I could have said more. I liked her face, her dark, intelligent eyes, and her friendly demeanor. But I’ve done this gig long enough to know the rules, both the ones Uber and Lyft lay down, and the ones we drivers develop for ourselves. I’ve learned to wait for cues from my passengers, to see how much they want to engage with me, and how familiar we allow ourselves to get while talking. I try to use an appropriate amount of discretion. But if a passenger wants to get friendly, I’m good with that.

Cassandra fumbled with her bag, finally got it open, and slipped her phone in.

“So, how’s your day going?” I asked.

“It was a beautiful day,” she said. “I did that walk along West Cliff Drive, from the wharf to Natural Bridges State Beach and back. I loved seeing all the people walking, jogging, and riding. Then I ended up at the bar at the Dream Inn. The people there were great. But really, the reason I came down from San Francisco on the bus today was to visit my grandfather. He’s in that senior care facility on Brommer Street, so that’s where we’re going.”

I drove us out of downtown, crossed the San Lorenzo River, and headed east. We cruised by the Santa Cruz Beach Boardwalk. The carnival lights glittered in the growing darkness.

“Tell me the coolest thing you’ve done today,” Cassandra said.

I drew a blank. I tried to think. There was nothing.

I laughed. “The coolest thing I’ve done today is give you a ride.” Yeah, it was jive and a little flirty. But really, the day had been pretty unremarkable up to that point. It was the best I could come up with.

She laughed. “Okay,” she said. “Let’s play Truth or Dare.”

To be continued.

Introduction

The Business End, Part 1

It's a strange gig. Uber calls its drivers "Partners." That's a crock. Uber looks forward to the day when we're all replaced by self-driving vehicles. I'm reminded of an old joke in the Special Ops community: "I must be a mushroom. All they do is feed me shit and keep me in the dark." Partner, mushroom, whatever.

Call me Ishmael, or Travis, or Chief, or Jaffy, or Johnny Rico, or William Mandella, or Lonesome Cowboy Burt. You can call me Pardner, or Pard, if you like. Or El Pardnerino if you're not into the whole brevity thing. But don't call me Partner. Don't insult my intelligence. Partners have a seat at the table. I have a seat in my Subaru.

These are stories of my adventures as a rideshare driver in Santa Cruz, California. When I say Santa Cruz, I'm really talking about Santa Cruz County. That includes the cities of Santa Cruz, Capitola, Soquel, Aptos, and Watsonville, as well as the numerous rural, unincorporated yet very distinct communities nearby. Occasionally I would take passengers "over the hill" into Silicon Valley or further north to San Francisco or even Oakland or Berkeley or Marin County.

All this really happened. My memory is not perfect, and while driving I had to pay attention to the apps, the roads, the traffic, pedestrians, cyclists, skateboarders, homeless people pushing carts with all their earthly belongings across the street, my passengers, and my car. I may have a few of the details wrong. I did not record any of the conversations. If there are factual errors, they are unintentional. I've tried to recall events and conversations as well as I could.

Names have been changed to protect the innocent. In my stories, everyone is innocent, if only by reason of insanity.

I also drove for Lyft during this time, but due to both geographic and economic factors, the vast majority of my trips were as an Uber driver. And really, distinctions between the two companies become less apparent each day. Over the course of my 4 years in Santa Cruz, I gave over 10,000 rides.

Most of these stories were written down the following morning while having a bagel and coffee at The Firefly Cafe in Santa Cruz. They were recorded chronologically in journal form. Thus, the reader may notice that much of it is written in the present tense.

But I have chosen to organize these stories into categories, while keeping some sense of chronology if it adds context. There is some categorical overlap in places. I've tried to deal with that as well as I could, while keeping the storyline moving along. I apologize for whatever confusion that may cause the reader. In the end, it all fits together, with some stories touching others, as all of life's adventures seem to be connected somehow.

In fairness I should point out that these tales are only a small fraction of the more than 10,000 rides I gave. Not every night is exciting. Not every passenger is memorable. But all this happened. People sometimes ask me to tell them my stories, so here they are.

Back to the Business End. A couple of years after I started writing these stories down, Uber announced its *Community Guidelines.* I've copied part of it verbatim from Uber's website as it appeared in November, 2017.

Here's the relevant bullet:

- Uber Policy on Physical Contact.
    - Physical contact with the driver or fellow riders.

      As our community guidelines make clear, you shouldn't touch or flirt with other people in the car. As a reminder, Uber has a *no sex* rule. That's no sexual conduct with drivers or fellow riders, no matter what. And you should never hit or otherwise hurt a driver or fellow passenger.

Up until then, there were numerous behaviors that could get a driver kicked off Uber's driver app, but nothing relating specifically to the behavior of passengers. The *Community Guidelines* changed all that. Except, nothing really changed at all.

It seems pretty much all of Uber's policies were thought up by corporate headquarters types who have never spent any time behind the wheel with the app on, picking up and dropping off real live people. Whoever those policy makers are, they're clueless.

If two passengers want to make out in the back seat while I'm driving them home, I'm not going to kick them out on the curb and report them to Uber. Hell, they're paying for the ride. Theoretically, all passengers are adults. If they want to play tonsil hockey and maybe cop a feel, well, that's what people do.

If somebody wants to make out with their driver, that's a different story—and definitely one worth telling.

I'll be your driver tonight.

> *I was a pilgrim on your shore, and you welcomed me*
> *With open arms and a smile*
> *And you said, "It ain't easy, but you could make it here,*
> *If you're willing to go that extra mile."*
>
> From the song *Goodnight, Santa Cruz*

I write songs, fiction, and tales that are mostly true. I write because I enjoy the process, the rhythm of a good sentence, and the magic of the fog in my brain becoming words on a page.

I'm the father of twin girls who are about to graduate from 2 different high schools, then go off to colleges somewhere. We came to Santa Cruz in search of education options for them that were better than those where we had been living, a southwestern state where public education is not a priority for the political party that controls both houses of the state legislature. We found some very good schools here.

I retired from the US Army as a Chief Warrant Officer, served as director of a local non-profit, then had my own small delivery company prior to moving to California. During much of that time, I've also been a singer/songwriter. I'm not a great musician, but I am a decent songwriter.

I'm divorced, but my former spouse and I have learned to work well together in the co-parenting project of raising our daughters. We all live in the same house in Santa Cruz because the rent is so high here, we can't afford two separate households.

I started driving for Uber shortly before we moved to California. I knew the job scene in Santa Cruz would be challenging. I saw being a rideshare driver as a good option for transitional employment. I've stayed with it because it pays the rent, which in Santa Cruz is not insignificant. Once we were settled in Santa Cruz, I signed on with Lyft, too.

I have occasionally joked that I started off doing this rideshare gig for the money, but I stayed for the adventure. That's not too far from the truth. Except every month the rent is due.

It's morning. I'm sitting at The Firefly Cafe in Santa Cruz, eating a bagel and drinking coffee. I drove until midnight last night. Gave 15 rides, made a hundred dollars. That's gross, not net. It's a hundred dollars before gas, oil, and maintenance on the vehicle, before insurance, before the weekly full-service car washes, the daily vacuuming at the self-serve car wash, before the bottled water and packets of Capri Sun I stock for the passengers, and before the bill for cell phone service with unlimited data. You do the math.

Maybe I should get out of here and do some driving. Like I said, every month the rent is due.

The Shape of the Land

Santa Cruz is located at the north end of Monterey Bay (this might be a good time to look at the map). That can, geographically speaking, be a bit confusing. While the Pacific Coast is generally to the west for most of California, in Santa Cruz, the ocean is to the south. Highway 1, which generally runs north-south along the California coast, runs west-east through Santa Cruz.

Sandwiched between the mountains and the sea, Santa Cruz is about five miles long from east to west, but only about 3 miles wide from the foothills to the beaches. The San Lorenzo River flows south through the center of Santa Cruz, entering Monterey Bay just east of the famous Santa Cruz Beach Boardwalk.

Downtown Santa Cruz is just west of the river. Ocean Street, the main north-south artery, begins at a levee near the Boardwalk, and heads north for about a mile and a half where it intersects Highway 1 and becomes Highway 17.

Highway 17 is a curvy and sometimes treacherous 4 lane road that winds generally northward through the Santa Cruz Mountains for about 20 miles until it drops into Silicon Valley where it eventually becomes Interstate 880 and continues north toward Oakland.

Surf City

In 1885 the Washington Monument was dedicated. The Statue of Liberty arrived in New York City. AT&T became a corporation. And Mark Twain published *The Adventures of Huckleberry Finn.*

Also in 1885, 3 young princes, David Kawananakoa, Edward Keli'iahonui, and Jonah Kuhio Kalaniana'ole, members of the Hawaiian royal family, who were attending a military school in San Mateo, heard there were some good waves to be found in Santa Cruz. They came south that summer, found a lumber mill that would cut some surfboards from redwood, and became the first surfers on the Pacific Coast of North America. One hundred forty years later, Santa Cruz is still Surf City.

Surfing is deeply embedded in the culture of Santa Cruz. On any given day, 4 generations of surfers will be in the water below West Cliff Drive. Kids as young as 10, old dudes pushing 70, and everybody in between are out on their boards, waiting their turn, waiting for that one perfect wave.

It's quite common to see vehicles going down the streets of Santa Cruz with wetsuits hanging from the side mirrors. That's what surfers do with their wetsuits when it's time to go home. It helps them dry out.

West of the Boardwalk are the two major (most popular) surfing areas in Santa Cruz. First is Cowell's, a broad stretch of beach just outside the doors of the

Dream Inn, where the waves tend to be fairly mild. It's where beginners start out. On a given afternoon, there might be as many as 50 beginning surfers laying on their boards in the calm water at Cowell's, hoping for a wave, any wave. It's a view that should be in the dictionary under "optimism."

Further west below West Cliff Drive, the biggest waves are found at Steamer Lane, which sits below a rocky outcropping that juts out into the Bay, creating bigger waves as the flow of the ocean wraps around it. It got its name when, during a time of unusually calm water, some surfers pooled their money and hired a steamer to run back and forth out beyond the point, hoping it would create enough wake to make some good waves. An old lighthouse at the point (still functioning) now serves as the Santa Cruz Surfing Museum.

In the classic 1960s surf film *Ride the Wild Surf,* a tale of the big waves at Waimea Bay on Oahu's North Shore, one of the characters is named Steamer Lane, an apparent nod to Santa Cruz and its surfing culture.

The University of California, Santa Cruz

Sitting on the slope of a broad hillside north of the west end of Santa Cruz, on land that was once used by pioneers for ranching and a limestone quarry, is the University of California, Santa Cruz. Born in the turbulent 1960s, UCSC reflects the mores and culture of the leftmost city on the Left Coast.

The school's mascot is the Banana Slug (check out the t-shirt John Travolta wore in *Pulp Fiction*). There is a strong history of student activism (Huey Newton, co-founder of the Black Panther Party, was an early UCSC graduate). There are sports teams, though no football team. UCSC is an NCAA Division III school, and remains independent of any athletic conference.

The school's unofficial motto is *A Tradition of Non-Traditionalism.* UCSC does have some interesting traditions. In the fall there's First Rain, a nude run through campus the night of the first hard rain of the fall quarter. And on April 20th each year (more commonly called 4/20), students gather in one of the campus meadows for a large (and officially unsanctioned) weed smoking event. *Rolling Stone* once called UCSC "the most stoned campus on earth," a distinction the university does not include in its official recruiting materials. They may, however, mention that UCSC's McHenry Library houses the official *Grateful Dead Archive Collection.*

The campus is large, much of it tucked in among a forest of redwoods. Organized somewhat like those revered English universities (Oxford, Cambridge), UCSC consists of 10 colleges, providing students with a sense of community within the larger university structure.

While there are fraternities and sororities, there is no Greek Row, per se, on or off campus. Some of the fraternities have houses in town around which they organize their activities. Most rideshare drivers are aware of these houses. There are parties at them nearly every weekend. We drivers frequently get called in *en masse* right after the cops show up, which happens right after the neighbors complain. The cops know where those party houses are, too.

I realize I've made UCSC sound like nothing more than a drugged-out party school. That really isn't a fair description. UCSC has some great academic programs, notably in technology and engineering, marine biology, astrophysics, and the social sciences.

Many UCSC graduates have gone on to do great things, benefitting from a liberal education on a beautiful campus. These include (but are certainly not limited to) astronauts Steven Hawley and Kathryn D. Sullivan, journalists Ezra Klein, Dana Priest, and David Talbot, film makers Cary Joji Fukunaga and Stephanie Allain, singer-songwriter Gillian Welch, authors Bell Hooks Jayne Ann Krentz, Lori R. King, Reyna Grande, and Amy Tan, actors Maya Rudolph, Mathew Grey Gubler, Camryn Manheim, Susan Ruttan, and Ally Walker, astronomers Nicholas B. Suntzeff, Geoffrey Marcy, Mark M. Phillips, and Tod R. Lauer, conductor/music director Kent Nagano, geophysicist Martha MacNutt, writer/producer Marti Noxon, and former YouTube CEO Susan Wojcicki. Silicon Valley legend holds that Google was founded in Wojcicki's Menlo Park garage.

While I have no doubt they took their studies at UCSC seriously, I imagine they had a little fun, too.

Chapter 1

Conversations, Part 1

One of the joys of being a driver is listening to the things people say. Sometimes it's just between them. Sometimes they lower their voices. Sometimes they include me in the conversation.

Jenny and Jack

During my first month of driving, I picked up a couple at the Safeway in Aptos. Let's call them Jenny and Jack. They weren't especially young, perhaps in their thirties. I don't recall how they ended up at the Safeway. But they got into the back seat with a cake and a few bottles of beverage in a brown shopping bag.

They told me they were newlyweds. I offered my congratulations. Then they clarified they didn't *just* get married, the wedding was actually 3 weeks before, but they were committed to behaving as if they were newlyweds on their honeymoon for an extended period of time.

They gave me directions to their house which was up a winding road somewhere above the Rio Del Mar Country Club area. About halfway into the drive, their conversation started to get a bit colorful.

"Honey, what do you want to do first when we get home?" Jenny asked. "Should we have dinner right away and then fuck, or should we tear each other's clothes off and just jump right into bed as soon as we get home, then eat whenever?"

Perhaps thinking about the bottles in the shopping bag, Jack said, "Well, all that sounds great, but I think I'd like a drink when we get home, you know, before we move on into the evening's activities."

"That sounds good," said Jenny. "Would you like me to give you a blow job while you're having a drink?" She said it as though it was something she would truly enjoy doing, not just a marital obligation.

At that point, Jack decided to include me in the conversation. "What do you think, Burt? Is it okay to sip a drink while your new wife is giving you a blow job? I'm sure she won't mind if I tell you she gives fantastic blowjobs."

This was certainly outside the scope of my usual driving conversations. But if the newlyweds were really seeking my counsel, I would do the best I could.

"Actually, that sounds like a great way to start the evening, Jack," I offered. "But in fairness, Jenny, I think you should have a glass, too. That way you could take a quick break now and then, have a sip, then continue. Maybe it will prolong the pleasure for both of you."

"Oh," said Jenny, "I like that idea. More fun for both of us. And I'll make sure I save a little in my glass, to help wash it down when he cums."

We got to their house near the end of the road at the top of the hill. Jenny handed me a $5 tip and said, "Thank you, Burt. I'm glad you were our driver tonight." I was grateful for the generous tip. That was before Uber added the option to tip through the app, and not every passenger thought to tip their driver.

Jack grabbed the bag of bottles while Jenny got the cake. They walked into their home, ready for a night of marital bliss. I backed my car out of their driveway and made my way back down the hill, trying to concentrate on driving instead of what Jenny and Jack were doing.

Daughters and the Emotionally Distant Daddy

One Saturday night I got a request for a pickup from a bar on The Esplanade in Capitola. The Esplanade is a two block semi-circle of bars and restaurants adjacent to Capitola Beach.

Not long after I signaled my arrival through the Uber app, two women came out of the Paradise Beach Grill. Once in the car, the one who had requested the ride, let's call her Connie, began sobbing. I noticed no one had entered a destination into the app.

"So, ladies, where are we going?" Her sister Julie gave me an address on Santa Cruz's West Side.

I pulled a pack of facial tissues out of the Subaru's console storage compartment and handed it to Connie. I think they were both surprised I had something like that close at hand. I've found they can come in quite handy.

Apparently their father, and perhaps other family members, had been in the bar.

"I can't believe I made a scene like that in the bar," said Connie, still sobbing.

"Oh honey, you didn't make a scene," Julie tried to reassure her.

"I don't know what happened," said Connie, still crying. "It was just so weird being there with Dad, and him pretending he really cares, even though he never calls just to say 'Hi' or ask how I'm doing. I mean never!"

"I don't think that means he doesn't care, Connie. That's just how he is. It's like he doesn't even know how to be close to anyone in the family."

They continued to dissect the elements of their family dynamics.

It's not always easy for me to pick up every detail in a conversation between passengers. But somewhere in the conversation, Julie said something about being on acid, like they both were.

I hope I got that wrong. Maybe the acid is different today than it was back in my reckless youth when Tim Scully was cranking out the Orange Sunshine at 350 micrograms per dose. Based on my (admittedly imperfect) memory, the last thing I'd want to do on acid is sit in some cheesy bar with emotionally distant family members. But hey, that's just me.

At some point as we were headed west on Highway 1, Connie said, "Burt, I'm so sorry if we made this a bad trip for you."

I replied, "No problem. I understand. My father never told me he loved me. I have two daughters, and I tell them that I love them every day." They didn't say anything in response, so I stopped talking.

As we were nearing the house that was their destination, Julie said, "Man, I hope there's some place here where we can smoke some weed."

I didn't get an especially good look at them when they got in the car, but they didn't look like a couple of stoner acid heads to me. But this is, after all, Santa Cruz, so who knows?

A Menage a Trois, or Else Not

I picked up two young women at a sushi place downtown. Let's call them Diane and Stacy. I'd guess they were in their mid-20s. The destination was Bocci's Cellar, a popular and somewhat rustic bar off River Street.

They were discussing a brief relationship Diane had with a guy. Their voices got kind of quiet, slightly above a whisper. Apparently the relationship with the guy didn't work out. Diane hadn't warned him she also liked girls. He freaked out a bit when they were at a popular downtown bar, and she danced with another girl.

"That chick Carol?" Stacy said. "I remember you dancing with her. She looked hot."

"She is hot."

I missed some of the details, but apparently Diane invited the guy to join her and Carol in a threesome. Stacy said, "Isn't that supposed to be every straight guy's fantasy, to be in bed with two girls?"

"I guess," Diane replied.

"Sounds like what really bothered him was you were going to get it on with Carol whether he wanted to join the party or not."

"And that's pretty much the way it ended," said Diane.

Taking Care of Business

He was tall, with a dark neatly-trimmed beard and dark hair. She had long, straight blond hair and was wearing a red dress with a high hemline. I picked them up at a restaurant in the hip and popular area near Capitola, around 41st Avenue and Portola. It was clear they'd had more than just a few drinks with dinner. They were going to an address off California Street, across town on Santa Cruz's West Side. So I drove up 41st to catch Highway 1 going west.

Before we got on the freeway, Boyfriend made a phone call. I always turn the music down once I know someone is on the phone, not so I can listen, but so they can hear. He was talking to a supplier, saying he needed some samples in order to generate some orders. I wasn't familiar with the product terminology. They could have been talking about kitchen tile or cloth shopping bags. But since this is Santa Cruz, it's more likely they were discussing cannabis products.

Somewhere in that conversation, Girlfriend started feeling either amorous, neglected, or both. She got close to boyfriend and began kissing his neck and face with considerable enthusiasm. Somehow he managed to conclude his first conversation, and then made a call to another supplier. Girlfriend wasn't fazed. She moved in closer, almost on top of him, as Highway 1 passed through the confluence with Highway 17 and transitioned from freeway to a 4-lane street.

He said, "Babe, I'm trying to take care of some business," but she was undeterred.

She continued to kiss and caress him quite passionately. "I want to take care of some business, too, baby."

Their two silhouettes in the rear-view mirror literally merged into one. At that point we were on city streets, going southwest on Mission. I was wondering if she might straddle his lap and do him right there in the back seat.

Fortunately, or not, he told me to turn left on Walnut Street. In a few seconds we took a right on California Street. Shortly thereafter we arrived at their home. Whatever additional romance was still on the evening's agenda would have to take place inside the house instead of the back seat of my car. Probably better for the seats that way.

Chapter 2

Ladies of the Night, Part 1

It seems I fall in love, or lust, at least once a week
I suffer serious bouts of infatuation, so to speak
In this beautiful town full of beautiful women
Surf City with a cannabis high
Still each night I wind up alone when my shift is done

From the song *Caroline*

There are aspects of this job that could make me crazy (and perhaps do). There is no shortage of beautiful women in this wild and crazy beach and college town. I give them rides to and from restaurants, parties, sorority soirees, the university campus, the bars, and, of course, home.

During the past year, the fashion look of choice, especially among the college women, is an elegant, skin-tight, black evening dress with a plunging neck and a hem so high it's just barely street legal. I'm actually very impressed with how they can find dresses that fit them so well, showing the absolute maximum amount of leg without revealing the color of their thongs. I inquired about it once and was told by one woman it requires some very serious shopping.

So I guess it's not surprising I should occasionally think about women while I'm driving, always wondering if the next encounter might transcend the usual driver-passenger relationship. It's like buying a ticket for Mega Millions or Powerball. It's a ticket to dream. From the moment I purchase a ticket until I read the winning numbers the next morning, I get to imagine how life would be with a cash flow of a few million a year—the trips I could take, the house I could buy, the charities I could fund, the college tuition I could afford.

When a ride request appears on the screen of my phone, and the name is Amanda, or Nicole, or Brooke, or Amy, or dozens of other female names, I have a ticket to dream. I get to dream until I pull up to the curb, and she and her boyfriend get in the car.

I've had some interesting encounters with women. I suppose it's the nature of the business. I drive at night. I pick up people who have been drinking or maybe smoking some weed, or both. Nearly half are women. Sometimes they get a bit uninhibited, especially if they're not out with a guy. Many are entirely forgettable. But a few have been quite memorable.

Savannah Red Pants

I got a request on September 10, 2016, just before sundown, for a pick-up in what's called The Village in the city of Capitola. The Village is about 4 square blocks of shops, restaurants, and bars in the old downtown, and is adjacent to Capitola Beach. There was a wine-tasting festival that weekend that took up about half of the Village, in particular the 2-block semi-circle known as The Esplanade. The requester's name was Savannah.

There is always a lot of traffic going through The Village. There are only a few roads in town that have bridges over Soquel Creek. One of those bridges is on the western edge of The Village. So The Village was jammed up with traffic, and some of the roads leading in and out of The Village, as well as the entire street through The Esplanade, were blocked off due to the wine tasting festival. The idea is to get people to park in parking lots a block or two outside The Village and have them walk to the festival.

Savannah sent me a text after I accepted her request for pick-up. She let me know it would be pointless to attempt to pick her up in The Village due to the traffic and road closures, so she and her friends were going to walk across the bridge and up Wharf Road which leads out of The Village. We texted back and forth while I was stuck in traffic. I ask what she is wearing, to make it easier to spot her among the throngs of folks walking up Wharf Road from the festival. She responded with "Red Pants."

I eventually found Savannah Red Pants and her two friends on Wharf Road. She got in the front passenger seat. The two guys with her got in the back. It's clear they had all tasted way more than their share of the wine. Simultaneously, they all tried to tell me the destination, which was an apartment near a popular restaurant just a couple miles away.

Savannah Red Pants told me one of the guys in back was her younger brother, and the other her ex-boyfriend. He'd been ex for 6 years or so. They called her "Vannah."

I can't give you a great description of Savannah Red Pants. I make an effort to not ogle my passengers. She had brown hair down past her shoulders. She had an attractive face and nice curves. She wore a loose-fitting white peasant top. Her arms were not the arms of a woman who spends a lot of time working out at a CrossFit gym; hard to tell about her age. She might have been in her early thirties. She was slurring her words a bit.

Savannah and I talked a little. She had me change my route by a few blocks to show me the house where she grew up. To the extent possible, we made friendly conversation over the din in the backseat.

The boys in the back were being loud, obnoxious, and rude. They were also drunk. They talked about women. One of them kept going on about wanting to find some ugly girl to fuck. There was a lot of talk about fucking, very loud talk about fucking. At one point one of them asked Savannah about her experience with girls. He pressed her, using graphic language. Several times Savannah replied, "I told you, I've never fucked a girl!" She was clearly not happy with this turn of the conversation.

Had the circumstances been different, I would have kicked both the guys out of my car after just a few blocks. But I wasn't going to kick Savannah out, so ejecting the other two would have been extremely awkward.

When we arrived at the apartment, they all got out. Savannah was about halfway out of the car. Her feet were on the ground, but her head and body were still in the car with me, causing her to bend over toward me. She said, "I just want to thank you for being so nice. I really appreciate it."

Clearly she was a bit embarrassed about the behavior of her brother and her ex-boyfriend. She held out her right hand. I took her hand in mine, but instead of shaking hands, we just sort of gazed at each other for several seconds while holding hands. If the situation hadn't been so awkward, I think she might have kissed me.

She got out, closed the door, and I drove off to the next adventure, wondering if I'd ever see Savannah Red Pants again.

Savannah's Little Sister

I don't know quite what to think about this next story. As Dave Barry says, "I'm not making this up."

Lately Uber has had a bonus system in place, rewarding drivers in some metropolitan areas for reaching certain milestones in terms of numbers of rides given each week. It's one way Uber tries to make up for the fact that they have cut rates numerous times in the past two years, and consequently drivers are making less per ride.

The bonus system works like this: If I give 30 rides in a week, I get an extra $50. If I give 50 rides, it's $100. If I give 80 rides, it's a $250 bonus. I've done 80 rides a few times. That's sort of become my standard for a good week.

By the end of Thursday night I had given nearly 52 rides. I could see that, with a good weekend, it was possible for me to do 100 rides. That would get me a $375 bonus. So I set that as a goal. I had a pretty good Friday night, racking up 20 trips.

My second ride on Saturday morning took me from Santa Cruz to downtown San Francisco. It was good in terms of how much I made, but spending that much time on a single ride put a dent in my plan to do 100 rides. I took a break when I got back to Santa Cruz, catching a nap and a shower so I'd be ready to go the distance on Saturday night.

I hit the streets around four thirty that afternoon, ready to go long into the night to try to reach that goal of 100 rides. I was prepared to push it into Sunday if I needed to, but I generally like to meet my goal on Saturday night, then take Sunday off.

Early in the evening I got a request for a pick-up at an address that looked familiar. That happens sometimes. That same week I had picked up the same girl at her house two nights in a row.

I arrived at the address and realized that's where I had dropped off Savannah and the boys a week before. But the name on the ride request wasn't Savannah, it was Tonya. A guy came out, followed by two young women. They all looked like they were in their twenties.

Once they got in the car, I asked if any of them knew Savannah. Tonya said, "Hell yeah, she's my sister. How do you know Vannah?" I said I had given her a ride from the wine festival the previous weekend. Tonya said something about the fruit not falling far from the tree. "She's out there somewhere tonight. Who knows, you might end up giving her a ride tonight, too."

Tonya's boyfriend was sitting in the passenger seat. He asked if it would be ok if we could go to the destination, get the wallet Tonya had left there, then return to the apartment. I said that would be fine if she could get the wallet real fast.

Then Tonya asked if we could stop at a bank with an ATM to get some cash. ATM stops usually go pretty fast, and people with cash sometimes give tips, so I said that would be okay. We talked about which bank she wanted to stop at. Then she and her friend and her boyfriend talked and decided they could hit the ATM at a 7-Eleven on Laurel, which is near the place where the wallet was. Tonya needed cigarettes. The boyfriend wanted beer. I reluctantly said we could do that if they could do it real fast. I'd been burned on store stops a few times. One went longer than 15 minutes. Time in the store is time I'm not meeting my goal of 100 rides.

They assured me it would only take a minute. We got to the 7-Eleven, they all got out of the car, I waited. Then I waited some more. Then I waited some more.

The boyfriend came out, apologized for the delay, and said he'd go back in the store to hurry the girls along. Finally, Tonya came out and said there was something wrong with the ATM in the store, and neither of the two employees could figure out how to make it work. The boyfriend said he was going to get the cash. Another five minutes passed. I was thinking seriously of terminating the ride and risking a bad review. Instead, I continued to wait. The fact that Tonya was Savannah's sister played a part in the decision, I'm sure.

Eventually they all came back out. Boyfriend had figured out how to work the ATM. We drove a few more blocks to where a guy was waiting to hand Tonya her wallet. Fortunately, that part went smoothly. We headed back to the apartment. On the way I gave Tonya one of my business cards, asking her to give it to Vannah so she could contact me if she ever needed a ride. Tonya took the card. "Hey, I might call you for a ride, too." I sincerely hoped not. The whole trip ate up 33 minutes.

Things did get better as the night went on, with lots of short rides, running up my numbers. At one fifteen in the morning I shut off the app. I had done 29 rides that day, giving me a total of 101 for the week. It would be a good payout from Uber the following Wednesday.

Kimberly

The following Monday night around ten I got a request for a pick-up at a Chinese restaurant in Capitola. When I arrived, a woman got in the front seat. Let's call her Kimberly. She had long dark hair. She was probably in her mid-to-late thirties. She was real perky and chatty, and she'd been drinking a bit.

I don't usually get a good look at my passengers, but we had a little time before her friends came out. Kimberly actually looks a lot like a documentary filmmaker I've known for several years. We made small talk while waiting for her two friends, a man and a woman. They eventually came out and got in the back seat. They were a couple. Kimberly was "third wheeling" it.

Along the way to drop the couple off, Kimberly and her friend talked about working out. I didn't catch all of the conversation. It was mostly about running. There was mention of running stairs. You don't hear much anymore about people running stairs, not since fitness centers were infused with those mechanical stair machines. When I was a young runner, we called it running stadium stairs. The best place to do it was a high school football stadium. Up the stairs, down the stairs. Repeat as needed. It was a great workout.

As a freshman in college, I lived in a third floor dorm room. I made it a habit to run up those stairs every time. Old habits don't break easily. All these years later, I don't walk up stairs, I jog up.

We dropped the couple off in the Pleasure Point neighborhood. Kimberly gave me directions to her house. It was a different route than the Uber app was giving me. Kimberly wanted us to go along the coast on East Cliff Drive back to Capitola, so that's the route I took. It was a beautiful night, with moonlight reflecting off the bay.

We must have discussed politics, though I don't really remember. It was an election year, after all. Trump had just been selected as the Republican nominee. And there were some GOP senate candidates saying really batshit crazy things about women and their bodies.

Apparently my progressive inclinations came out. I guess Kimberly liked whatever I said. At some point in the conversation she put her hand on my shoulder and said, "You're the kind of guy chicks just love."

I probably should have asked if that included her. Instead, I just said, "Thank you. You're very sweet."

During our conversation she was turned a bit toward me. She reached out, putting her hand on my upper arm or shoulder three or four times during our ride together, not like she was hitting on me, more as a way to put a physical accent on what she was saying.

In an effort to be professional, I couldn't really say, "That feels really nice. Touch me some more." But that's definitely what I was feeling.

I pulled up in front of her little cottage, on a bluff above The Village. A navy blue Volvo SUV was parked near the gate. We had arrived at her destination.

That often leads to that somewhat awkward moment in which it feels like we've connected, bonded, become friends, but the ride is officially over, and we may never see one another again.

We each said nice things, wishing one another all the best. Kimberly undid her seatbelt. Then before getting out of the car, she turned, reached out and leaned over to give me a hug. In the process of embracing, we ended up kissing one another on the cheek. The hug lasted a few seconds.

Eventually ending the hug, she released me, said, "Good night, Burt," opened her car door and got out. She walked up to a wooden gate and pushed on it a couple times, but it didn't open. I watched as she took a half step back, then delivered a perfect front snap kick, Tae Kwon Do style. The gate burst open, Kimberly walked through the gate, and disappeared into the night.

Not to put too fine a point on it, but I need to say it. In over a year as a rideshare driver, no passenger had ever given me a hug before. Ever. But it wouldn't be the last time.

A Little Closure - Sort Of

About six months after my encounter with Kimberly, I was driving down a Westside street on my way to the gym. As I pulled up to an intersection to make a right turn, I saw her car, the navy blue Volvo SUV with no roof rack, going in the other direction. I caught a glimpse of her profile as we passed. It was definitely Kimberly.

I made my turn, then when I could I made a U-turn, then a left turn, heading back where I had come from. There was no sign of her ahead of me. I drove down the street, looking on each side of the street to see if perhaps she had pulled in somewhere. When I got to the end of the block I looked left and right for her car. I couldn't see it anywhere. So I made another U-turn which got me back in my original direction of travel.

Just to make sure I hadn't missed something, I pulled into a parking lot near where I first saw her Volvo. It's a spot near a popular bakery and yoga studio. I saw her car parked there.

I parked nearby. As I was getting out of my car, I saw her come out of one of the businesses and walk toward her car. I walked over toward her as she was getting to her car. She looked apprehensive, perhaps thinking I was homeless and about to ask her for money. It's probably the beard.

I said, "Hi. Forgive me for bothering you, but I was your Uber driver about 6 months ago."

She looked at me somewhat incredulously and said, "You were?" Not the response I was hoping for. It was clear she didn't recognize me, and was still probably apprehensive about even speaking to me.

"Yes." With the absence of any recognition, I wasn't sure what to say. All I came up with was, "I just wanted to let you know you were one of the nicest passengers I've ever had. So, thanks. That's all." I turned and walked back to my car.

Perhaps that rattled around in her brain for a while. Maybe she will remember. Maybe she won't. Maybe she'll wonder how I even remembered what she looked like after all this time. I would.

And now a little postscript. I've driven down Kimberly's street a few times, either picking up or dropping off passengers in her neighborhood. Recently I noticed her Volvo finally has a rack on top. It's there to hold her surfboard. Kimberly has joined the ranks of Santa Cruz's surfer chicks. I think that's awesome.

Chapter 3

Lost and Found, Part 1

People leave things in my car. I have found various articles of clothing, groceries, jewelry, glasses, beer and wine, phones, wallets, umbrellas, calculators, and makeup products. Sometimes I have been able to return these things. Other times not.

One morning, a few days after Valentine's Day, I was restocking the passenger doors with water and juice packs (yes, I'm one of those drivers who does that). At the bottom of the cylindrical space where I was about to place a water bottle, I noticed a small rectangular shaped item. I picked it up. It was an unopened condom. I examined it carefully. No love note. No phone number. It was clearly not a Valentine left for me by an amorous passenger. Who knows how or why it got left there. I replaced it with a bottle of water.

On a busy night, Uber's big computer will have matched me with my next passenger before I have arrived at the destination of my current passenger. As soon as I drop off my passengers, there are more waiting for me somewhere. This doesn't leave much time to take a look at the condition of the back seat. I may have a dozen different passengers before I get to check out how my back seat looks. Things that are dark may go unnoticed on the floorboard for more than an hour at night. By the time I, or perhaps a passenger, notice an item, I might have no idea who left it behind.

If I think I know who might have left an item, I can contact them through the app by letting Uber or

Lyft know I found something. Occasionally, I've been able to return to the house where I dropped the passengers off to return the item. I have even posted a few items in the lost-and-found section of Craig's List when I could think of no other way to make contact.

Sometimes a passenger will notice something is missing, and if they think it might be in my car, they can contact me by going through the company (Uber or Lyft).

The funny thing is, most passengers don't seem to notice they're missing anything.

I'll confess to maybe being overly conscientious about returning items left in my car. Some might even say "compulsive." I can't really explain that compulsion, except to say that I'd want someone to do the same for me.

In a couple of these Lost and Found stories, the former Intelligence Analyst in me seems to show up wanting to solve the mystery of who an item belongs to, and how I can get it back to them. If I occasionally went above and beyond what most drivers would do, I'm okay with that.

The Big Tipper

There's a place I park sometimes while waiting for a ride request. It's a little plaza on the corner of 17th and Brommer. There's a gas station/convenience store right on the corner. A small parking lot separates the gas station from a tiny strip mall, a contiguous L-shaped building of commercial spaces with one end on 17th and the other end on Brommer. Part of the building has two floors. There's a wide, outside stairway leading from the edge of the parking lot to a balcony on the second floor.

People's Coffee, a nice neighborhood coffee cafe, is at the Brommer end of the businesses. I frequently get a mocha there early in the evening to carry me into the night. At the other end there is a gym that takes up most of the spaces, upstairs and down, from 17th to past the bend in the L. The gym is part of a local chain of gyms. I work out at the one on the Westside.

Early on weeknights there is a very popular Zumba class in the gym. Sometimes it's hard to find a parking space. On a Thursday night I managed to find the only available space, right in front of People's Coffee. I backed my car into the space, got my mocha, and waited in my car for a ride request.

Over by the gym I saw something I had never noticed before. There were two women running up and down the stairs. One of them had long, dark hair. I wondered if she might be Kimberly. This was not long after I had given her a ride home, the one in which we ended up hugging each other before she got out of my car.

I thought about walking over to the stairway, but I couldn't think of a good way to interrupt two women in the middle of their workout without appearing to be a real jerk. Then I got a ride request and had to go.

That request was from someone named James. The pick-up spot was a hotel on Portola, a street which runs east-west near the coastline, connecting Santa Cruz and Capitola. James was a short, Irish looking guy, not stocky, but with the broad shoulders and strong arms that made me think he's spent many years doing some kind of construction work. There was a younger couple with him. They might have been his daughter and her husband, or maybe it was the other way around. At any rate, they seemed like family.

I think James had already gotten started on the festivities before I arrived. He was jovial in a joking sort of way, though I actually couldn't make sense of most of what he said to me. At one point, we passed a corner on Portola that had a large vacant building on it. James said something about developing it together with the young man. He asked me what I thought the building would become. I guessed, "Warehouse?" He just laughed. Apparently that wasn't it, though he never said what it had been or what it might become. Months later I learned they were turning it into condos.

I took them to Shadowbrook, a very popular upscale restaurant on the banks of Soquel Creek. When we arrived, James acted like he really wanted to give me a tip; wanted to be generous and appreciative, but neither he nor the couple could come up with any cash (this was before Uber changed its policy and added a tipping option through the app). He apologized profusely. I told him "No worries," and went on my way.

My next passenger was a guy I'll call Ned. He needed a pick-up at a place where he gets his dialysis treatment three times a week. He struggled physically coming out of the building, using a rolling walker with a few bags piled on it, stuff he needed during the hours of dialysis he had to undergo nearly every other day.

I placed his stuff in the back of my car, helped him into the front passenger seat, and folded the walker up so it would fit on the back seat. As I drove, Ned told me a little about his dialysis routine. "It's like a job you go to every couple of days. Except, with a job you get vacations. With dialysis, you never get a vacation." I let that one sink in.

His house was in one of the many communities up in the forested hills above Santa Cruz. We arrived at his home. I got his walker and his bags out of the back. As Ned was getting out of the car, he noticed a cell phone on the seat where he had been sitting. He thought it was mine, but mine was still in the bracket on the dashboard where I keep it for music, navigation, and the rideshare apps.

I thanked him and stuck it in my pocket, telling him I thought I knew to whom it might belong. I shut off the Uber app on my phone so I wouldn't get any more requests, then hurried back to Capitola.

The Shadowbrook parking lot sits along Wharf Road, on a high point above the western bank of Soquel Creek. The restaurant itself is down below, near the creek. There's actually a gondola that takes the customers down to the restaurant from the parking lot, then back up again when they're ready to leave. There are also some very nice stairways that go down to the restaurant. I took the stairs.

Once inside the restaurant I explained to the maître d' what my mission was. He checked his list. Nobody named James was on the list. But he was kind enough to tell me I could look around the bar and restaurant to see if I could find them. I found them sitting at the bar.

James recognized me as I walked up to him. I asked if he had his cell phone with him. He said, "Yeah, I think I had it right here on the bar somewhere." He started looking around. He couldn't find it. I held out the phone Ned had found on the seat. "Any chance this is yours? Someone found it in my car."

James took the phone, pushed a couple buttons and said, "Yeah, that's mine alright. Thank you. Here, let me give you a tip." Then he realized he and his ride mates still had no cash. Again, he apologized profusely. He said he wanted to give me a hug. So I hugged him. Then he said, "I have a confession to make."

I could only imagine what he needed to confess. He said, "I'm going to vote for Trump." James, the small-time developer is a fan of Trump, the big-time developer.

I have no recollection of saying anything that might have revealed my political leanings. Perhaps I'm more transparent than I realize. Maybe it's the beard and ponytail.

I said, "Enjoy the rest of your evening, James." Leaving the restaurant, I jogged up several flights of stairs to get back to the parking lot. It was a good workout.

Red Red Wine

For reasons I don't understand, Main Street in Soquel is not the main street in Soquel. That honor goes to Soquel Drive. Soquel Drive actually begins in downtown Santa Cruz as Soquel Avenue. It goes east for about three miles until it crosses Highway 1 diagonally, then continues east, paralleling Highway 1, as Soquel Drive. It goes through downtown Soquel, then gradually curves south, staying within a few blocks of Highway 1, through the Cabrillo College campus, and into the town of Aptos.

Then there's Porter Street. Porter begins as Bay Avenue in Capitola (since it was the road from Soquel to Capitola Bay, I guess). Bay goes north out of the Capitola Village, and becomes Porter once it crosses under Highway 1. A few blocks further north, Porter becomes the Soquel-San Jose Road, essentially the old highway through the Santa Cruz Mountains, linking San Jose to the east side of Santa Cruz, much the same way Highway 9 does a few miles to the west. That was before Highway 17 was carved through the mountains.

The intersection of Soquel Drive and Porter Street is the hub of the four or five blocks that pass for a downtown in Soquel. Main Street begins at an intersection with Porter just a block north of Highway 1. It goes east for about 100 yards, then curves left and goes north, eventually crossing Soquel Drive and continuing north into the foothills of the Santa Cruz Mountains.

Along that short little section of Main Street just off of Porter there are four dining and drinking businesses. On the right are Carpo's, which is sort of a family restaurant with a salad bar, and Beer Thirty, an afternoon and evening beer joint popular with the young adult crowd. On the left-hand side are Michael's on Main, a drinking and dining establishment with outdoor seating down by Soquel Creek, and which features some of the best live music in the area, and Little Tampico, a nice little Mexican restaurant and bar. They are all busy pretty much every night.

Late one Saturday night I responded to a request for a pick-up at Michael's on Main. There I picked up two couples who had been dining and drinking. As they got in, I heard the clinking of what sounded like a couple of wine bottles. I figured they bought a couple bottles to take with them when they left Michael's. I almost had it right.

I took them to what I assume was an Airbnb on Opal Cliff, near the coast in Capitola. They seemed more like weekend tourists than locals. They were quite jovial and a bit loud on the trip there.

I dropped them off right in front of the house, drove a little bit further up the street and made a U-turn. As I did, I heard a strange sound coming from behind my seat. Then I smelled it.

I pulled over, got out, opened the left rear door and looked in. They had apparently left behind a bottle of red wine on the floorboard. It must have been opened earlier that evening, then plugged with a cork. When I made my U-turn it fell over, the cork popped out, and the left side of my rear floorboard was drowning in red wine.

I turned the bottle upright. There was still a little in the bottle that hadn't spilled out, but not much. I picked up the cork and stuffed it back into the bottle. I found a rag in the back of my car, and used it to soak up as much of the spilled wine as I could.

After cleaning the floorboard so it was just moist instead of sopping wet, I took the bottle and the rag, and walked over to the low wooden fence in front of my passengers' Airbnb. I put the bottle atop the fence post nearest the driveway, and hung the rag on the top rail next to it. I imagined in the morning they would come out and find the rag and the bottle. Maybe they'd figure out what had happened, or maybe they'd just assume I drank most of their wine.

At that point I decided I was done for the night. I drove away with the sweet aroma of red wine wafting through my Subaru. There are worse smells. I've probably encountered most of them on one trip or another.

Chapter 4

Conversations, Part 2

Sometimes You Get Drama

I was driving on a Thursday night in mid-October. It was slower than usual. The tourist season was definitely over.

It seemed like there were more Uber drivers working than there had been the previous three nights. Back then, we thought we could get a sense of how many drivers were working by opening the passenger app. It would show locations for drivers in the area who are not currently committed to a ride. On a slow night, more drivers show up on that screen because there are fewer requests, thus more drivers are available at any given moment.

But like much of Uber, it turned out all that was just an illusion, or perhaps more accurately, a con job. The passenger app was designed to give passengers the sense there were plenty of drivers just waiting to give them rides. But those drivers weren't really there.

I tested this out one Sunday morning while parked near the UCSC campus. The passenger app showed 3 additional drivers within a block of where I was parked. Out of curiosity, I did a little drive around the block. There were no other cars displaying the Uber decal, and no drivers just sitting behind the wheel of their parked cars as though they had nothing better to do on a Sunday morning.

So, who knows if there were really more drivers out that Thursday night. All I knew for sure was I hadn't been getting many requests.

It was a little before nine that night. I had only given four rides since starting at around five that afternoon. I was parked in Capitola, in the Village near the Esplanade, mostly because no other drivers were hanging out in that area, when the request came in. The address was only a few blocks away. It was Kimberly's street. I wondered if it might be her. What are the chances?

The name of the requester was Michelle. Now in truth, I don't actually recall Kimberly's real name. She told me when she got in the car that night over a month before, but I really can't remember what she said her name was. I just call her Kimberly to give her a name. Maybe her name was really Michelle.

But when I got up the hill and on her street, I realized the address was not Kimberly's address. I drove past Kimberly's Volvo, then continued another couple of blocks up the street. On my way there I got a text from Michelle. Here's what it said:

Michelle: Hello. I'm not going to be riding with you. I just need you to deliver something to me. I live in Oakland. And I just need some glass blowing supplies delivered to me from a friend's house in Capitola. Just making sure you're willing to drive to Oakland.

I replied, "Ok."

Requests like this are unusual, but not unheard of. I've done a couple pick-ups at a pharmacy and delivered them to a senior care facility. I've talked to one other driver in Santa Cruz who has done one of those, too. I also once had a request to pick up a dinner order downtown and deliver it to a house a few blocks away. That was, of course, before Uber Eats and all its clones became a thing.

I pulled up in front of the house. It was pretty dark, with no streetlights nearby. It's an area with no sidewalks, and no real curbs, either. There was a spot in front of the house that was the only logical parking place on the street. There was a sign in the yard adjacent to the space. It said *Please park somewhere else.* I didn't see anywhere else to park, but I wasn't planning on parking. I just wanted to get the supplies and be on my way. It should have been easy. It wasn't.

The front of the house was U-shaped. There was a door on the left and one on the right. They looked like different residences. Through a window in the door on the left, I could see a guy watching TV. I knocked on the door. He answered. I told him Michelle had sent me to pick up some glass blowing supplies. He had no idea who Michelle was or what I was talking about. He said I could try the other two residences. Neal lived to the right and Donna lived in the back.

I went back to my car, and sent Michelle a text.

Me: Who's my point of contact here?

Michelle: She doesn't have good service. I am trying to get a hold of her now. Let me know when you're there.

Me: I'm parked out front now. Is it Donna?

Michelle: Yes. How did you know?

Are you talking to her now?

She's not answering her phone. It may have just died. She may need a minute to plug it in.

Me: Talked to her neighbor. The house is a triplex.

Michelle: Ok. Give me a minute. Hopefully I will get a hold of her in the next minute or so.

She was supposed to drop this off herself. But she is in pain for some reason and can't make it. Her phone just died. You might have to knock on her door. Her service is making this a pain in the butt.

I can cancel if you have to wait more than a few minutes.

At that point it looked like I was going to be there longer than I originally thought. So, I turned on my emergency flashers (common practice when parked to make a pick-up), made my way around the side of the house in the dark, and located the back door. I knocked on the door. Donna answered the door saying, "Who are you?"

"I'm the guy Michelle sent to pick up the glassblowing supplies."

"I have no idea what you're talking about. Please leave me alone." She seemed to be a little bit out of it.

I apologized, went back to my car and began to text Michelle. As I was nearly done with my text, my phone rang. It was Michelle. I told her about my exchange with Donna, including that I thought she might be somewhat medicated. Michelle said she still wasn't having any luck reaching Donna on the phone.

I was thinking it might be a good time to cancel the request when Donna appeared at my car door.

"Please turn your flashers off. It's very indiscreet."

"Okay. Can I get the supplies for Michelle."

"Yes. Just please turn off your flashers. It's so indiscreet." Apparently I might be attracting some undesired attention. I turned off my flashers. She turned and headed around the house to the back yard.

I called Michelle to let her know I had made contact with Donna again and she said she would bring out the supplies. Then I waited.

Michelle: Thank you very much. If you do see her before I talk to her, let her know she can keep the $40 she owes me for doing me this favor.

Eventually Donna reappeared. She said she had trouble finding her car keys. She mentioned she was drunk. She unlocked her car and pulled out some kind of tank of pressurized gas. She told me to place it valve-side down. So I put it on the floor of my back seat, angled so the valve was down. I thanked her for her help, and told her Michelle said she could keep the forty dollars. She replied with something about keeping it, but on her own terms.

Back in my car, I sent another text to Michelle.

Me: Ok. I have 1 tank. Was that all?

Michelle: There should be two.

She said she had two for me.

If you're with her, can you tell her to call me?

Me: Already told her. She's a bit drunk.

At that point I went to the back of the house and knocked on Donna's door again. It sounded as though she was talking to someone. I wondered if she had connected with Michelle. She said, "Come in."

I walked in. There was no one else there. The room, which probably was her living room, was somewhat in disarray. There was a single scuba diving tank just inside the door, slightly smaller than the one she gave me to take to Michelle. Just to my right was a small bathroom. The door was open. Donna was sitting there peeing. She was not on the phone. Perhaps she had been talking to herself.

I apologized. It was an awkward moment, at least for me. It didn't seem to faze Donna much at all. I told her Michelle said I was supposed to pick up two tanks. Donna seemed confused. I called Michelle again, and for a while I was a conduit for a conversation between Donna and Michelle until I finally reached into the bathroom and handed my phone to Donna who was still sitting on the toilet with her pants down, so she and Michelle could speak directly to one another.

Overall, the conversation didn't make much sense to me, as it involved whatever agreements and history existed between the two of them. Donna mentioned being in some pain in her ovaries, that she was considering going to the ER.

Eventually they concluded their conversation. Donna stood up, handed the phone back to me, pulled up her pants and flushed the toilet. She said she would get the other tank. I asked her if she would like me to drop her at the ER on my way to Oakland (Highway 1 goes right past the hospital). She said, "No," looking at me like it was a bizarre suggestion.

I returned to the front of the property. Donna followed a minute later. She unlocked her car again, pulled out the second tank and handed it to me. I put it in my car.

As I closed the backseat door, Donna walked up to me. Apparently personal space isn't a factor in her computations. I'd say she got in my face, but she was considerably shorter than me, so her face was about 6 inches from my chest. She started telling me that because of me having my flashers on, she was being kicked out of her house. She laid it on pretty thick, "Well I hope you're happy. Because of you, I'm being kicked out of my house and will have nowhere to live." She repeated that a few times, guilt-tripping me as well as she could in her drunken condition. I apologized, reminding her I was just an Uber driver, responding to a request the best I could.

Then she said, "How about you give me that fat sixty dollar tip you're going to get?" I said I hadn't heard anything about any tip. All I knew was, I was driving the tanks to Oakland.

Once again, I apologized for any problems I may have caused, and I thanked her for her assistance.

I got in the car, called Michelle and told her I was on my way. She said, "I'll see you in about an hour and a half."

It took a little longer than that to get there. Michelle's house is actually up in the hills above Oakland, up a winding road where you have to watch out for the deer that graze on the grass on the side of the road. I lost the cell phone and data signal just a block from Michelle's house. There were no street lights, and addresses were impossible to read. I had to drive out along a street that overlooks the city to get a good enough signal to call Michelle to get her exact location. It was a little after 11 when I handed her the two tanks.

There was no sixty dollar tip. There was no tip at all. But it was an eighty dollar fare for me. So, while I only gave five rides for the night, I made over a hundred dollars. Some nights you just have to take what you can get. Some nights you get drama.

Strippers

One night I got a text from a woman named Daisy. She said she got my card from a friend. She wanted to know if I could take her to San Jose. Daisy gave me her home address which was just outside of Watsonville.

She came out of her house wrapped in a heavy coat and carrying a small suitcase. She was a bit on the petite side, so at first I thought she was younger than her actual age. She had dark hair and Latina features, and spoke good English with a bit of a Spanish accent.

When I asked where we were going she said, "Do you know where the Gold Club is?" I plugged the name into the Waze app, and said, "I've never been there, but I know where it is." The Gold Club is San Jose's premier "gentlemen's club." Daisy is a "dancer."

I did this trip a few times over the next couple of weeks. Sometimes I'd pick her up in San Jose around two in the morning to bring her back home. Other times she'd let me know she had a ride. She paid me in cash, and she paid me well.

It got a little more interesting one evening when I drove to her house to pick her up for a trip to San Francisco. Her friend, a co-worker, was coming along. Let's call her Brandi.

Daisy came out of her house to tell me we had to wait a few minutes for her friend. Brandi pulled up in her late model Camaro a few minutes later. Brandi is a tall blonde with a rather striking figure. That is to say, her boobs are larger than one would expect to see on such a slender frame. Better living through cosmetic surgery.

Daisy and Brandi got in the back seat, and we headed north for San Francisco. The ladies were on their way to audition at a club in The City.

Their conversations were unremarkable, yet somehow fascinating. At first, the talk focused on the manager of a club where they both had worked, and what they didn't like about her management style. They also talked about "the business end." Apparently entertainers like Daisy and Brandi are independent contractors, much like rideshare drivers. That means no benefits (health insurance, unemployment insurance, workers comp, paid time off, etc.).

Brandi is on probation. DUI, I think. So she's in a program and has to go to meetings. Part of her deal is she must show she is working (thus allowing her to drive), but she can't have a job that involves serving alcohol. That limits her options.

She talked about how she got into the business. Her husband had been injured and wasn't working. They have a couple of kids. Brandi was good looking, liked to dance, and was a little wild anyway. She told her husband she thought she could make some money as a "dancer." He asked if she thought she could make two hundred dollars a night. She just laughed. Apparently on a good night she makes considerably more than that. When she showed her husband the cash from her first night's tips, he was down for it.

Brandi said, "I figured I'd just do it for a few months, maybe a year at the most, just to get us out of debt, maybe get a nice car."

"That's how it starts for everyone," said Daisy.

"Yep. Pretty soon I was paying for a boob job, bariatric surgery, other cosmetic surgery. Now I'm starting to think about another boob job."

"I know," said Daisy. "It's not easy to look eighteen when you're getting close to thirty."

"Tell me about it. Especially after having a couple kids."

Both women have children. Daisy has a daughter who lives with Daisy's mother. Brandi and her husband have two kids at home. Both tell the kids they do something else for a living. Daisy says she's a waitress. Brandi says she's a bartender. Neither are happy about lying to their children, but they also don't want their children to know how they really earn their living. Brandi said she gets really pissed off when one of her friends fails to be discreet about her occupation when her kids are around.

The conversation stayed pretty mundane like that until we pulled up near the club in San Francisco. There was a quick discussion between the women about paying me. Daisy had cash but Brandi did not. Brandi was hoping I used Square so she could just run her credit card. At that time, I had not yet set up an account with Square.

Brandi said, "Is there some other way we could take care of this? How 'bout if we show you our boobs?" She looked as if she was about to pull down the front of her tank top.

"Brandi," Daisy said, "remember what I said about keeping it professional with the driver."

"Oh yeah. Right. Sorry."

I ended up with a partial payment. I had done enough business with Daisy to know she would get the cash from Brandi at the end of the night, and give it to me the next time. I made a mental note to sign up for Square.

With a Little Help from Her Friends

On a Saturday morning I started driving right after dropping one of my daughters off at school to take the PSAT. I was in a pretty good position numerically, having completed 66 trips through Friday night and sitting at $515 earnings for the week before any bonus money. I knew I could easily get to eighty trips, earning the $225 bonus. With a really big ride count on Saturday, I'd be willing to go back out on Sunday if it looked like I had a shot at a hundred trips (which would make the bonus $350). That was during the first couple of years of driving, when making a grand a week was actually possible.

My 5th ride on Saturday morning was a pick-up on the UCSC campus. It turned out she was a student at UC Berkeley who had been visiting a friend at UCSC. At 10:45 we took off for Berkeley, an 81 mile trip.

Because of the rain, there were multiple accidents on Highway 17 that Saturday. There were places where the severed exterior body parts of various automobiles were scattered along the shoulder of the highway. It's a challenging drive for some people on a normal day. Add some rain and it can get messy.

So it wasn't the fastest trip I've ever made to the East Bay. I caught lunch at the Rubio's in Emeryville on the way back. Made it back to Santa Cruz about 3:30 that afternoon. I definitely wasn't going to get 100 rides for the week, but the $86 payout on the Berkeley trip (and $81 trip to Oakland a couple of nights before) would help make up the difference.

After a nap and a shower, I went back out on the street. By then, the rain had arrived in force and visibility sucked pretty bad. Around five, I received a request for a pick-up at an afternoon college party at an apartment complex not far from downtown Santa Cruz. The party was just breaking up as I arrived. There was another Uber driver who must have arrived just before me.

I'd been waiting there for a couple of minutes when a girl came out and asked me if I was there to pick up Emmy. When I said I was, she said, "That's not me. I'll go tell her you're here." I waited for a couple more minutes.

Eventually a group of kids came out. Emmy was drunk and hanging on to some guy. The party was over, but she didn't want to leave this guy she just met. Her friends were trying their best to get her into the car. Finally they succeeded. Her friend Shonda got her buckled safely into the back seat. Shonda rode up front in the passenger seat. Another friend got in the back seat next to Emmy.

Emmy was pissed and vocal about not wanting to leave yet, going on about how she finally met a cute guy and then had to leave before she got his number.

Shonda was a pretty sharp black girl from LA. She told me Emmy started drinking and didn't stop. "That's kind of her pattern," she said. "But all the girls were leaving, and I wasn't about to leave her there drunk with a house full of guys."

Since Emmy was too drunk to tell me her address, Shonda gave me directions to her house. It was in an area on the Westside, downhill from the university, populated mostly with students.

"You're a good friend, Shonda," I said, "even if she doesn't appreciate it at the moment." As we were getting near the house where Emmy lived, Shonda said she would like to take an Uber home, but she was having trouble getting the Uber system to change her account over to her new credit card. I asked Shonda where she lived. She gave me an address on Broadway, about a mile east of downtown. I said, “I'll get you home.”

Once we arrived at the house where Emmy lived, I closed out the ride, then logged off the Uber app. Emmy's housemate got out of the back seat, then helped Emmy get out. Together they made it in the front door of the house.

I took Shonda home, off the app, no charge. After the way she made sure Emmy got home safely, I figured it was the least I could do.

The 7-Hour Lunch Date From Hell

I'll be the first to admit that I don't understand modern dating. While I have dated a bit since getting divorced several years ago, I still think of myself as somewhat relationship-challenged. But it wasn't until I heard this story that I realized how little I really understood.

I picked up a couple in Capitola, at a house just off the upscale commercial area around 41st Avenue and Portola Drive. Let's call him David; we'll call her Sandy. They were headed into Santa Cruz on a dinner date.

When I asked the generic "How's your day going?" question, Sandy confided that she had just gotten back from the worst imaginable lunch date, one that lasted for seven hours. She called it "The Lunch Date from Hell."

As a driver who likes good conversation, and who is also a bit curious about the lives of my passengers, I couldn't just let that one drop. So I asked Sandy how she ended up on the 7-hour lunch date from hell.

"Well, basically, this guy, his name is Don, asked me to lunch. It was a day off for me and I had some time. He said he wanted to take me to this place in La Honda."

"He wanted to drive all the way to La Honda for lunch?" I asked, incredulously.

Here I should interject a little geography. La Honda is an unincorporated community (you can't really call it a town) with a population of about 1,000.

It's located in the Santa Cruz Mountains between Highway 1, which follows the rugged Pacific Coastline sort of northwest of Santa Cruz, and the foothills that form the western edge of Silicon Valley. You get there by driving 40 miles up Highway 1 from Capitola to San Gregorio, then going about 8 miles on Highway 84, also known as La Honda Road, which winds east through mountainous terrain to La Honda. There are other routes, but that's the shortest one.

All that is to say La Honda is the sort of place you aren't likely to pass through on your way to somewhere else. Most people only end up in La Honda if La Honda is their destination. I had only heard of it because it is mentioned briefly in Tom Wolfe's "*The Electric Kool-Aid Acid Test.*" From Capitola, it can take a couple of hours to get there, depending on traffic.

"Yes," said Sandy. "He said there was this really cool place he wanted to take me to have lunch. It seemed a little odd. But it was my second date with him and the first date was nice, so I thought, what the heck?"

"So you went all the way to La Honda on a lunch date?" I asked, still trying to wrap my head around the 7-hour lunch date from hell.

"Yes. I figured it would be a pretty drive, you know, along the coastline and then up into the mountains. And it was pretty. But there I was next to him in his pickup truck, and he was pretty much feeling me up the whole way there. The lunch place wasn't bad. Not great, but not bad. Then it was back down La Honda Road in his truck, and more of the same."

"That sounds really creepy."

"Well, let's just say I felt like I needed a really good shower when I got home. And if I had it to do over, it's not the way I would spend my day off."

"So, how did you meet this guy?"

"David and I were at Suda a couple weeks ago. We were at the bar waiting for a table. Don was sitting next to me at the bar. We talked a little. He asked for my number. Then he called me a few days later and we went out. Like I said, the first date was nice. I guess you can't always tell from a first date."

"So, David," I asked, "you were there with Sandy at Suda when she was chatting with Don?"

"Yeah," said David. "He seemed like a nice enough guy."

After dropping Sandy and David at a Sushi place in Santa Cruz, I tried to imagine myself on a date with a girl, and, while we're at a bar, she's giving out her number to another guy. I decided David must be a whole lot more secure than I am when it comes to dating relationships. Either that, or I don't have a clue how dating works these days. Probably both.

Chapter 5

Uber Therapy, Part 1

Credit Where Credit is Due

One evening, I was giving a ride to a young woman I picked up in Scotts Valley. Let's call her Nancy. She said, "I want to give you something. I don't want anything for it. I just want you to give me credit for coming up with the idea. Since you drivers are in such a great position to listen to people telling you about all their problems, I call it *Uber Therapy*. You can feel free to use that term, just please give me credit for coming up with it."

I promised Nancy I would give her credit. Unfortunately, while I'm remembering to give her credit, I don't remember Nancy's real name. Names disappear from the app as soon as the ride ends.

Once we established that I was her therapist for the time it took to drive from Scotts Valley to the home of a friend in Santa Cruz, Nancy commenced to tell me what a jerk her ex-husband was, how she had assumed responsibility for raising their child, and how she was a much better parent than her former spouse.

This was apparently her first night out in quite some time, largely because of her ex taking so little responsibility for his share of parenting their child. She was getting together with several girlfriends, the sort of social event she had missed out on for some time.

I don't think she was actually looking for therapy that evening, but I gave her props for being such a good, responsible parent. I think that was what she really wanted to hear.

She would probably do well to address the resentments she has toward her ex. As somebody told me once, "Holding on to resentments is like drinking poison hoping the other person will die." But that's not the sort of thing we could cover in a ten minute Uber ride.

I'm grateful she gave me the term *Uber Therapy*. There have been a few times when it has seemed quite appropriate.

Lenny

On a Saturday morning in September, my first month of driving in Santa Cruz, I responded to a request for a pick-up near the corner of 17th Avenue and Portola. The app said the passenger's name was Lenny. It took me a few minutes to find him. He was walking along 17th, coming up from the beach. All he was wearing was some board shorts and a pair of tennis shoes. He had his t-shirt in one hand and a six pack of beer in the other.

Lenny was a fairly buff dude. I'd guess he was younger than 30, average height and appeared to be in pretty good shape. His light brown hair was trimmed somewhat short for Santa Cruz. He had multiple tatts on both arms.

Lenny got in, I confirmed he was my passenger, hit start on the Uber app, and discovered we were going over the hill to San Jose. We'd have a lot of timc to talk. Lenny liked to talk.

By the time we were heading north on Highway 17, I'd learned that Lenny was a veteran. He left the Army as a staff sergeant. In 8 years he'd done tours in both Iraq and Afghanistan.

Both out of curiosity, and as a way of establishing my own military bona fides, I asked him, "What was your MOS?" MOS stands for Military Occupational Skill. Only someone who has served would use the term "MOS."

"I was a scout," he said. "Mech Infantry."

"How long have you been back?"

"Four, no, more like five months."

"I'm glad you made it back. What are you doing now?"

"I've got a construction crew. We do remodels, some additions. Mostly residential. Then I've got another gig I do at night."

"Really? What's that?"

"Male stripper. There's a club I work at a couple nights a week."

"You like it?"

"Shit man, I love it. Women sticking cash in my jockstrap. Telling me how hot I look. I make more in two nights at the club than I make the rest of the week doing construction."

"Cool. So what's happening in San Jose?"

"There's a girl. I've known her for a few years. All that time, we've just been friends. But it seems like we sort of like each other, you know, like more than just friends. So I'm going up there to see her and to see if it's really going anywhere."

"And you think she feels the same way about you?"

"I think so. That's why I'm going to see her, to see if she's into it, too. Hey, is it okay if I have a beer while we're driving there?"

"No Lenny, that is not okay. If I get busted for an open container violation, then I'm not working anymore, and that's not okay. I got kids at home. I gotta pay the rent and put food on the table."

"Okay. I get it. How about vaping? Can I vape?"

"No. But I'll tell you what. Once we get to the summit, I'll pull over. You can get out, drink a beer, and vape. Then when you're done, get back in and we will proceed over the hill. Will that work for you?"

"Yeah, that'll work. What did you do in the army?"

"Me? I worked in intelligence. Spent some time in Special Operations. So Airborne, Jumpmaster, Pathfinder. I retired as a CW3."

"No shit? A Chief Warrant Officer? I guess that explains it."

"Explains what?"

"Why you won't cut me any slack. But I get it."

About then we arrived at the summit, a flat area with a store on one side of the highway and a cafe on the other side.

"Here we are, Lenny. Enjoy."

Lenny got out, pulled a can from the plastic 6-pack holder, popped the tab, and began to drink. I think he only brought thc can to his mouth three times before he'd drained it. Then he fired up his vaping device, took a couple of hits, exhaled the last of the electronic smoke, stuck it in the pocket of his shorts, slipped into the passenger seat, and we were back on the highway.

As we hit Los Gatos, the first town on the north side of the mountains, I noticed Lenny was starting to tense up a bit. He wasn't saying much, not like the talkative guy I started out with.

About the time we hit the San Jose city limits, Lenny said, "Man, I'm starting to think maybe this isn't such a good idea."

"What's going on, Lenny?"

"Ah, man, I'm starting to freak out. I mean, I don't even know what I should say to her?"

"How long have you known her?"

"A few years. I'd see her sometimes when I was home on leave, and I've seen her at a couple of parties since I've been back. But fuck, man, seriously, maybe this isn't a good idea."

"Does she know you're coming today? Is she expecting to see you?"

"Yeah, we talked on the phone, so she knows. But I just don't know how I'm supposed to do this."

I could tell from the GPS map on my phone we were just a couple of miles from her apartment. Lenny knew it, too.

"Fuck, man. Stop! Turn around! Let's go back. I can't do this!"

"Lenny, listen to me. Guys like you and me, we know how to do the hard stuff. We're good at it and we thrive on it. But we don't have experience doing the soft stuff, so it's scary. I get that. Here's the funny part. All you have to do is tell her you're a little nervous, that you're not sure how to do this, not sure what you're supposed to say. But you really do like her, and you want to spend time with her. Do you think you can say that?"

"You mean just tell her this is kind of scary for me, that I'm not sure what to say, but that I really do like her?"

"Yes. Just like that. That's all there is to it. For about a half second you're going to feel really vulnerable. But here's the thing. She's going to like you more for being honest and telling her what you're really feeling."

"You really think so?"

I pulled into the parking lot of her apartment complex and stopped the car.

"Absolutely. All you have to do is march up to that door, ring the doorbell, and tell her exactly what you just told me. Do you think you can do that?"

"Yeah, Chief. I think I can. Yeah. Thanks, man."

I held out my hand. As we shook hands I said, "Good luck, brother."

"Thanks, Chief. I really appreciate it."

I watched Lenny put his t-shirt on, walk up to the door, and ring the doorbell. When the door opened, she must have invited him in. I ended the ride and headed back over the hill. Mission accomplished.

Chapter 6

Ladies of the Night, Part 2

Body Language

I picked up a woman downtown late one Friday afternoon in September. Let's call her Sally. When she got into the passenger seat, I realized I had given her a ride a few weeks before—a round trip from her house to a downtown drugstore, then back again. She had been sick, with strep, I think, but said she wasn't contagious anymore. I'm not sure what she got at the drugstore--maybe medicine, maybe cheap booze.

I want to be kind in my description of her. I'll start by saying Sally is not my type, whatever that is. While she is likely chronologically younger than me, years of hard living have taken their toll. Her voice is rough with the sound of whiskey and cigarettes. She has a quality that conveys that, even if she was clean and sober, she hasn't been for very long.

After she got in the passenger seat, I said, "I think I gave you a ride a couple of weeks ago. CVS and back. You were just getting over being sick."

Sally didn't say much, barely a response, as if that was something she wished nobody remembered, and if she didn't talk about it, maybe nobody would. I said, "I hope you're feeling better." Again, no response from Sally.

We rolled north on River Street toward her place, a little low rent location on the other side of Highway 1 (or at least what passes for "low rent" in Santa Cruz). It's on an obscure little road that wanders into the woods but doesn't really go anywhere.

As we were going down River Street, Sally slouched in the seat like she was relaxing. Then she did this move that's similar to the one teenage boys often do on a first date, sort of yawning with arms spread out from the shoulders, then one hand surreptitiously winds up around the girl's back and on her shoulder.

But Sally wasn't putting her hand on my shoulder. She put it on my thigh.

I didn't react. I just kept on driving, doing my best to pay attention to the road while a strange woman had her hand on my thigh. Not down by my knee, either. Her hand was high on my thigh, with her fingers resting on the inside of my thigh, very casually as though we were old friends with benefits for whom this level of intimacy happens all the time.

After maybe 30 seconds she moved her hand off. After a few more seconds, she put her hand right back on my thigh. There was a message in that, I'm sure. She was saying, *You might have thought that first time was just a careless accident. Now you can be sure it wasn't.* Clearly Sally wasn't much into talking, but her non-verbal communication was unusually straight-forward.

I should be clear, Sally wasn't fondling me. She wasn't moving her hand up and down my thigh, the way I have occasionally with a woman on a date, knowing she's enjoying this touching as a prelude to even more touching later in the evening. Sally wasn't doing that. She was just leaving her hand on my upper thigh, perhaps letting me know the next move was mine if I was interested.

I turned off River Street at the road that led to her house. The road starts out fairly smooth, but gets a bit rough as it makes its way into a more rural area where her house is. Sally took her hand away as the road got a bit bumpy. Her hand just bouncing around on my thigh could have gotten a little awkward. No telling where it might have landed.

I turned in on the dirt road that is essentially her driveway, stopping in front of her humble abode.

As she got out, she thanked me for the ride. There was a kindness, even sort of a sweetness in the way she said it. I wondered if she was really thanking me for not saying anything hurtful, not questioning her motives, or just not rejecting her outright. She took a bit of a risk. There wasn't really any payoff for her, but hopefully no emotional damage, either.

I drove back toward River Street. The sun hadn't even set yet, and I couldn't help but wonder what other strange adventures this wild and crazy town held in store.

Wild Birds and Chocolate

On a Friday night in November, I drove until around one in the morning. One of my last rides was a strange one. I picked up a woman in a mobile home park off of 7th Avenue. Let's call her Jennifer. She wanted to know if I was okay taking her to the liquor store on Seabright and back again. I agreed. She hopped in the front passenger seat. It was late, she was my 21st ride of the day, and I was probably going to call it a night very soon.

Jen was mildly attractive. I would have guessed she was in her 40s. I would have been wrong.

She began telling me about her new housemate who has a bunch of tropical birds living with her in Jen's mobile home—uncaged, apparently. The birds kept attacking Jen, scratching her hands and arms. She went on about this, all the way to the store. I tried to be supportive.

I parked in front of the store. She realized she didn't have her driver's license. She hoped they didn't card her. I told her they might since she looked pretty young. She laughed and said she was almost 60. I said, "No way!"

"Way," she replied. "I'm 58."

She went into the store. I waited. She was only in there for a couple of minutes. When she came out she had a bottle of Bacardi and a Toblerone chocolate bar. As we drove back toward her place she apologized for not having any cash for a tip. "How about some chocolate?" she said.

"That would be great," I replied. It was late. I could use a little chocolate boost.

She unwrapped an end of the chocolate bar, and offered me some. I was driving, and I didn't want to get chocolate all over my hands and the steering wheel. I asked if she would break off one of those triangular pieces and place it in my mouth. She did. She did it three times. It was about as intimate an experience as I'd had since the night I took Kimberly home. I wanted to hold her hand while I sucked the chocolate off her fingers—but I didn't.

As we were pulling into the mobile home park, she said she would really like to get a boyfriend, but clearly that wasn't going to happen with a house full of aggressive birds. I knew that was certainly true for me.

Sweet Confusion Under the Moonlight

Marx (Karl, not Groucho) said everything happens twice; the first time as tragedy and second time as farce. I'd argue that the first time doesn't necessarily have to be tragic. It could be quite nice. But I think he got that second time right.

It was a weeknight in mid-September, nearly a year after my encounter with Kimberly. The University students hadn't yet returned for the fall quarter, and the summer tourist season was winding down, so business was a little slow.

I was downtown. It was a little past dusk. I got a request for a pick-up in front of The Blue Lagoon, a downtown Santa Cruz night spot that is sort of a sports bar, but often has live music and dancing, at least on weekends.

Let's call her Carlene. She was fairly tall, had long, dark hair, and was friendly. She sat in the front seat, and started off by telling me that she was both stoned and a little drunk. She said it was her night off, so she came into town to hang out, drink, smoke weed, and enjoy the evening before having to go back to work the next day. She worked as a waitress downtown.

While I didn't recognize Carlene, she was certain I had taken her and her neighbor home within the past couple of weeks. I granted that it was possible, though I didn't recall. She went on to tell me several things she remembered from our previous conversation, none of which sounded accurate. But we had a nice chat just the same. At some point she mentioned she was married.

She lived several miles out of town, up Highway 1 toward Half Moon Bay. She pointed out her turnoff, and we followed a rough road for about another half mile. At a fork, she said, "Now you remember, you dropped my friend off at her house which is down that road, then you brought me home on this road." We followed the road another quarter mile until pulling up in front of her place. I'm certain I had never been there before.

It was dark in front of her house. Carlene apologized for not having any cash left with which to tip me. Then before getting out, she turned and gave me a hug. It was a nice hug. It was also a flashback to the hug from Kimberly the previous year. Only this time I was aware there might be a husband somewhere watching us through a window. So I didn't let the hug last very long, and there were certainly no kisses.

Even so, I find myself swinging past The Blue Lagoon on weeknights now and then, just in case Carlene needs another ride.

Chapter 7

The Business End, Part 2

The Rating Game

People often ask me what it's like being a rideshare driver.

I say, "Well, you know how on a regular job you have a boss, usually a manager or supervisor. And periodically, maybe every 6 months, you and your boss sit down together and have *the talk.* You discuss how well you met your goals and objectives, how well you worked as part of the team, stuff like that. And to a large extent, how that talk goes may have an impact on whether or not you get a raise or a promotion. Now imagine if you have that talk a dozen times a day, and half the people you talk with are drunk. That's what it's like being a rideshare driver."

Most rideshare passengers know that drivers are rated on a scale of 1 to 5 stars, with 5 being the highest. What most passengers don't know is, from the perspective of Uber and Lyft, a rating of 4 means something wasn't right, and below a 4 indicates some serious problem.

Drivers have a cumulative rating that passengers can see. With Uber it's an average of a driver's previous 500 rides.

Here's the killer: If a driver's cumulative rating falls below 4.6, the driver is suspended, and must get additional training before using the driver app again. That can create significant stress for drivers, especially new drivers who are prone to making rookie mistakes.

Another little-known fact: For the past few years, the annual turnover rate of drivers is around 50 percent. A more recent estimate is that over 60 percent of drivers decide to do something else within their first 6 months. That means there is a good chance you will get a driver who is not especially experienced. [1]

For most of my first year, my rating hovered between 4.85 and 4.88. That's not bad for a rookie, but it also indicates I made a few stupid mistakes early on. And apparently I make fewer stupid mistakes now. For the past year, my rating has consistently been in the 4.95 to 4.97 range. Both Uber and Lyft seem to be happy with that since it is well above average for active drivers at any given time.

A source of frustration for drivers is that a passenger might give a lower rating based on things that are not under the control of the driver. These can include slow traffic, detours, and surge prices.

Surge prices are increases in the cost of rides when demand significantly exceeds supply in the general area around the pick-up location. Uber and Lyft let passengers know, prior to completing the ride request, that surge prices are in effect. But alas, not all passengers read the fine print or do the math.

Drivers aren't told which passengers give what ratings. But sometimes it's not too difficult to figure that out. Here are a couple of examples.

During the third week of September the students returned to the University and the parties began. My last request on Friday night was for a pick-up at a party at an apartment complex near the university. Two young men who looked like college students got in, one of them carrying a couple of bottles, one of which was open. I asked what they had. He told me. It was something I'd never heard of. He said, "The chicks love it."

I told them I could not transport them if they had an open container. One of them said, "Why not?"

---

[1] Written in 2019

"Simple," I said. "That's 'go to jail' shit, and I lose my job. Not gonna happen." They acted as if nobody had ever told them what they could and could not do before, especially some Uber driver. It was clear they were irritated, but I have no reluctance transitioning from "some Uber driver" to "The Chief" when the situation requires it. They reluctantly agreed to keep the lids on the bottles, but they weren't happy.

As we were pulling away, a car pulled up. Three girls got out. They were all wearing tight t-shirts and very short shorts. I stopped and said to the college boys, "You guys might be leaving a little too soon." They took two seconds to check out the girls and agreed.

One of them said, "Go ahead and end the trip. I'll make sure you get your whole fare," like he could magically manipulate the Uber app to make that happen. Frankly, I was happy to have them out of my car. I'm also sure they gave me a very low rating. My rating average, which for months had bounced between 4.86 and 4.87, dropped from 4.87 to 4.84 that night. That score is based on the average of my last 500 rides, so it would take a "1" (on a scale of 5) to move my average that much. I decided to exercise more discretion in giving rides to college boys at known party locations.

Drivers get to rate passengers, too, and drivers get to see those ratings when a request comes across the app. A low passenger rating plus a college party location is now sufficient reason for me to cancel a ride once I see the pick-up location (currently, a driver sees the passenger's rating before accepting the ride, but doesn't see the pick-up location until after accepting). While Lyft doesn't care for such tactics,

Uber doesn't mind so long as the driver's cancellation rate stays below 10% of the number of rides accepted.

Saturday, October 29th, 2016, was essentially the 2nd Day of a 4-day Halloween weekend. Halloween is a major thing in Santa Cruz, a bigger deal for adults than it is for kids. There are lots of costume parties, and thousands of people end up downtown on Halloween night. It's such a big deal, the city cordons off much of central downtown so there are no cars on Pacific Avenue, just lots of people. There is a significant police presence, as well. Imagine a few thousand people in costume who have consumed some combination of booze, weed, and who knows what other recreational chemicals, roaming the downtown streets in an atmosphere of freaky holiday. That's Halloween in Santa Cruz. I guess it's as close to Mardi Gras as we get.

Saturday was a kick-ass day. I made over $300 (gross) on Saturday. My first ride of the day was a trip over the hill to an upscale neighborhood near Redwood City. That was good for $55.

There was some surge money during the night that pushed my dollar total up high. My last ride was a kid I picked up at a party on West Cliff Drive, and drove home several miles up into the hills north of Santa Cruz. He might have been a college student, but he looked to me like he was probably still in high school. I wondered if his account was linked to the credit card of one of his parents. Maybe he was new to using the app and didn't realize he had signed on for a 4.5 surge (meaning it cost roughly four and a half times as much as a non-surge trip). I made $85 on the ride, so it must have cost him (or his parent) over $110, with Uber taking the difference. By the end of the night my rating had dropped from 4.86 to 4.85. I suspect someone expressed displeasure with the cost of the ride by trashing my rating. That happens sometimes. It's the breaks of the game.

There's a popular blog for rideshare drivers called *The Rideshare Guy*. It was started by a driver who does a good job of staying on top of trends in the gig economy, and shares tips about everything from gadgets and apps for drivers to class action lawsuits filed by drivers against Uber or Lyft.

One of the gadgets he frequently recommends is a dash cam, especially one which can record what the driver sees on the road as well as what's happening inside the vehicle. Dash cams are frequently mentioned any time there's a news story about an assault on a driver, or when a passenger makes accusations about a driver's conduct. Dash cams can be very handy in providing evidence as to what actually happened.

I installed a dash cam for those very same reasons. I had been using it for less than two weeks when I picked up a young woman. I think she was a college student. To be legally compliant with the appropriate laws, I had a notice posted on the dashboard letting passengers know their ride was being recorded.

The young woman looked at the notice, then at the camera and said, "Is that recording right now?"

I said, "Yes it is. It's for your safety and mine."

She replied, "Is that even legal? I haven't given you consent to video me. Turn it off."

"It's legal as long as I advise passengers that the dash cam is operating."

"I don't care. Turn it off. I find it very intrusive, and I do not consent to you making a video of me."

I could tell there was no point in further discussion. I flipped the switch, turning the dash cam off. Shortly thereafter we arrived at her destination.

The next morning when I checked my rating, it had dropped from 4.87 to 4.85. I began to reconsider whether I really needed that dash cam.

Technical Difficulties

Sometimes technology changes everything.

I never really gave it much thought, but until the shift from 3G to 4G cell phone communication, the rideshare industry wasn't even a possibility. Before that, you could call a cab company, catch a bus, or take the subway or Metro or BART or MARTA or whatever they call it where you are.

Ironically, 4G was initially a response to the demand to stream music. But some innovators saw how other apps could exploit the increased speed and volume of data a 4G network could provide. Uber and Lyft were right there at the right moment in history, and could envision a revolution, not only in how people requested transportation, but how an entire industry could be leveraged through the use of an app on a smartphone.

All that new technology is absolutely marvelous. It brought drivers and riders together with just the touch of an icon on a smartphone. The heavy hand of the rideshare company, the owner of the app, controlling what passengers are charged and what drivers are paid, is almost invisible.

Rideshare drivers are at the mercy of the company apps we use. And let's face it, technology, for all its wonderfulness, isn't perfect. Sometimes there's a problem with the app, or perhaps with the big computer that interfaces with the app. When an app is hosed, we're hosed. There was definitely some hosing going on this past weekend.

I got a request for a pick-up on Laurel Street. The address and the GPS said the pick-up location was the 7-Eleven, the same one where Savannah's little sister Tonya had such a hard time getting money out of the ATM on the previous Saturday night. I arrived, looked around, didn't see a passenger, so I parked in the parking lot with my flashers on. After a couple of minutes with no contact, I sent a text to the ride requester, saying I was in the parking lot of the 7-Eleven. A minute later I got a reply that essentially said, "Why are you there? We're waiting at the (such-and-such) restaurant." It was 6 blocks away. By the time I arrived in front of the restaurant, the requester had canceled the request, no doubt frustrated by the delay.

The same thing happened two more times. On the first of those two, the requester called me in response to my text. She and her date were at a restaurant about a half mile from the pick-up address on my Uber screen. I got there in just about another minute. They were nice folks, and understanding when I said the same thing had occurred just a few minutes before.

The last time it happened, the request came from the UCSC campus. The pick-up location on my screen was the campus trailer park. This is an obscure and out of the way little area most students don't even know about. I only knew it because I'd been up there once before.

The topography of the campus is such that cell phone voice and data connections are frequently lost at several locations on and around the campus. This is because the campus sits on a large piece of land that rises above Santa Cruz on the northwest edge of the city, but also slopes away from Santa Cruz to the west, essentially shielding much of the campus from cell phone coverage. It's ironic that the UC campus nearest Silicon Valley, the technology hub of America, has the worst wireless connectivity of any university in California, but that is the case.

I got a call from the prospective passenger, alerting me that he was not where I was looking for him. Moments later, I lost both the Uber app and navigation. I tried to find him, but had no luck. I went back down the hill and off campus, rebooted my phone a couple of times, and finally got the app back up and running again. Once I got back to campus, I found my passenger.

When the technology fails, it can feel like Uber is intentionally trying to waste my time. Twice on Saturday night, as I was pushing really hard to get my numbers up to a respectable weekly total in order to get a decent bonus, I got sent on boondoggles.

The first was an address on Kennedy. I thought the pick-up was at one of the condos where I'd been a few times. It's in an area off Highway 1 near the border between Capitola and Aptos. I followed the GPS beyond the condo area, then looping back toward Highway 1, I turned on a street that paralleled the highway, and finally dead-ended in a little commercial zone I'd never seen before. The address was one of four in a two-story building. It was some kind of winetasting room. The place was closed. Inside there was a guy cleaning up.

I waited in front, sent a text saying I was in the parking lot, then waited some more. I called the number. I got the voicemail of the guy who owned the tasting room. Finally I went to the door. A guy opened the door somewhat reluctantly. I said I was an Uber driver, and told him the name of the supposed passenger. The clean-up guy had never heard of him. I canceled the ride as a "Rider No Show." I got paid $3.75 for wasting 20 minutes, counting travel time.

The second boondoggle was an address supposedly up a mountain road in a very rural area between the town of Scotts Valley (north of Santa Cruz) and the town of Felton (one of a series of communities in the San Lorenzo River Valley north of Santa Cruz). The entrance to the road had several Private Road and No Trespassing and No Outlet signs. Very welcoming. The GPS map showed I would have to go quite a way up that road.

I followed the road as it curved up the mountain. The terrain put me outside of phone and data coverage. The only light was my headlights. Little driveways broke off to the left and right at various spots along the road. I saw a sign that seemed to imply the road would lead to Hihn Road, a road I knew that makes a big loop off Highway 9 near the town of Ben Lomond north of Felton.

Not long after that I came to a fork in the road. At each fork there was an automated metal gate. There had been no contact from the requesting passenger telling me there was a gate and giving me the code. I managed to find a space wide enough for a 3-point turn-around. When I got back in data range, I canceled the ride as a "Bad Address."

That was enough fun for one night. I had spent nearly an hour on two phantom requests. My back hurt from driving most of the day and night. I went home, knowing I would have to drive on Sunday to make my quota.

On Sunday night I gave my 80th ride of the week. At around 9:30 p.m. I got the following email from Uber:

For a short period this past weekend, we experienced technical issues that might have affected your ability to complete trips through the partner app. Issues like this do not meet the high standards we set for ourselves, and we're sorry for the trouble this may have caused your business.

We've fixed the problem and have given you an additional $8.79 for the impact it had on your ability to use the platform. Everyone was affected differently, but we've determined this amount based on your individual experience during this time. You will see this reflected on your next payment statement.

Hard to know how they computed that $8.79 number. It's the equivalent of two fairly short rides. Still, it's an acknowledgment that the app is not as spectacular as Uber often likes to pretend. In fairness, let's give kudos to Uber for at least trying to do the right thing. And hey, $8.79 will almost cover a trip to Taco Bell with my girls.

Two weeks later I got another update from Uber about "technical difficulties."

Last Friday night there was a technical error with the app that may have affected your ability to take trips and qualify for promotions. Because of this, we lowered trip requirements for last week's promotion and credited accounts for any missed earnings, which will be reflected on your next pay statement.

During Friday's outage we were not able to uphold our standard of reliability. Please know that we have fixed the error and are working to ensure this doesn't happen again. Thank you for understanding and let us know if you have any questions.

-Your Bay Area Uber Team

I went back into my trip history for Friday. There were two trips where connectivity broke down. One I reported as a No Show (for which I get paid $3.75). The other was a trip in which, about a quarter mile into the trip, the route line (which is supposed to represent the route actually driven), branched into a V. It looks like the left branch followed the actual, logical route toward the pickup location near the UC campus, turned toward the destination, then disappeared. The right fork followed what would have been a near straight line from the pickup to the destination, an "as the crow flies" route.

I'm guessing somewhere on that trip I lost the data link. So Uber used the "as the crow flies" distance to supplement the actual (but only partially tracked) route from pick-up to destination. Guess I wasn't the only one who had app issues on Friday night. The problem is, due to the topography of the region and the poor wireless data coverage, Friday wasn't really much different than any other night.

I had two other events that represented inexplicable technology failures. One of them is the beginning of the story *The Kindness of Strangers* in Chapter 21.

The other took me to a hilltop above The Mystery Spot, an iconic Santa Cruz tourist attraction in the forested hills off Briarcliff Drive north of Santa Cruz. The GPS map took me through the Mystery Spot parking lot. At the end of the parking lot there is a gate with a push button keypad. Fortunately, I had the code because a semi-regular client lived in a cabin just a little way beyond the gate.

The route on the app had me continue further up Mystery Spot Road. Due to the mountainous terrain, I soon lost my data link. I continued up the road, as it seemed to be heading toward the passenger pick-up point.

I arrived at a relatively flat high point at the top of the hill. With no nearby ridgelines or mountain tops blocking the signal, I was able to reestablish my connection to the Uber app. The GPS feature said I was right where I was supposed to find my passenger.

There were a couple of houses up on that hilltop. I waited a minute to see if anyone came out. When nobody did, I got out of my car, walked to the nearest front door and knocked. When a man answered the door, I explained I was an Uber driver, and asked if anyone there had placed a request for a ride. He told me no one there had, and he was sure nobody was home at the other house, though I could knock on that door if I wanted to be sure.

I thanked him, walked to the other door and knocked. He was right. Nobody was home.

Since I had a data link, I canceled the request as a bad address, and headed back down the hill and into town.

Based on input from my tech-savvy daughters, not long after these episodes, I changed my cell phone carrier, and switched my phone from an Android to an iPhone. I think both have improved my coverage on the UCSC campus and perhaps out in some of the more remote regions, as well. So, it's nice to make a little technological progress, even when it's not perfect.

Chapter 8

Hurlers, Part I

My Uber Baptism

Drivers call it *The Uber Baptism* the first time it happens. It's the question about self-driving cars nobody has answered satisfactorily. It's when somebody vomits all over the seats of your car.

Mine occurred late on a Saturday night. Actually, more like one o'clock on Sunday morning. Admittedly, I was tired, sort of running on autopilot, and not giving the situation my full attention. I had only been a rideshare driver for a few months at that time.

The request was for a pick-up in an apartment area near the Boardwalk. Students rent there if they can't get something nicer (but more expensive) closer to the university.

It was dark on the street. I pulled up slowly, then stopped, looking around trying to find the address. While I was stopped, a group of students came out. I probably should have got out of the car to assess the situation, but I didn't. While I was talking to someone on the driver's side, trying to identify the person who placed the request, a girl got into the back seat on the other side. I didn't see how drunk she was. They had been drinking and eating pizza.

Eventually, I had two passengers in the back and one in the front. We headed for a dorm at the university.

About halfway there, I noticed the girl in the back slouching down with her head against the window. I had not yet learned that passengers who pass out are often the ones who also get sick.

We were actually on campus and nearing the dorm when she sat up and got sick. She hurled on the seats and the floorboard.

I pulled into the circle in front of her dorm. Everyone got out. She hurled again on the side of the car. I quickly assessed the mess on the seats and floorboard. She was starting to waver, so I took her hand. She assured me she was fine and could stand by herself. I let go. She fell down. I suspected she was going to feel pretty lousy the next day.

The guys who were with her didn't live in her dorm. There was talk about how to get her up to her dorm room. One of them wanted to call her roommate to come down. She said her roommate was out of town for the weekend. I was concerned about her being in her room alone, passing out again and maybe drowning in her own puke. Eventually, somebody who lived in her dorm agreed to help her up to her room and alert one of the RAs on her floor.

I took photos of the mess in the back seat, then cleaned up the semi-solid pieces (probably pizza) as well as I could.

I submitted a report to Uber through the app, attaching a few of the photos. Obviously I was done for the night. Nobody was going to want to ride in my car for a while.

Uber reviewed my claim, approved it, letting me know I would get an extra $200 for that pay period to cover my cleaning cost. It would be charged to the account of the guy who put in the request. I felt a little bad for him. Hopefully he was able to work something out with his friend once she was functional again.

When I got up on Sunday morning it was raining—not real hard, just enough that my usual car wash was not open. I drove around town a bit. None of the car washes were open.

By Monday the sky had cleared. I got to the car wash and paid for the interior cleaning as well as a regular car wash. The whole thing cost me $75. Then my seats were damp for the rest of the day, so I couldn't drive on Monday, day or night. So I guess the $125 that remained covered my down time.

With my lesson learned, I bought a box of small garbage bags, the size you could use in your bathroom trash can. I put some in the pockets on the back side of both front seats, easily accessible from the back seats. It was a smart move. They have come in handy on numerous occasions.

The 2-Bagger

It was after midnight on a Saturday night. I was heading into The Village when I got a request for a pick-up at JJ's Cantina in Soquel. There I picked up a woman we'll call Elena. As she was getting into the passenger seat, I asked her how she was doing. "I'm a little drunk," she said.

Shortly after we got on Highway 1, she asked if I had any water. "There's some on the door next to you," I replied. She picked up the water bottle and took a few drinks.

Her destination was Watsonville. She wasn't making much conversation. As we headed south on the freeway, she laid her head back as though she were going to take a nap. I've come to recognize that as a warning sign. Passengers who look like they are taking a nap may well be passing out. Passengers who pass out may also throw up.

I thought of reaching behind her to grab a bag from the pocket behind her passenger seat, but I didn't want to freak her out with some unexpected move in her direction. I should have listened to my gut.

Moments later she sat up, put her hands to her mouth and said, "I think I'm going to be sick."

I quickly grabbed a bag from behind her seat and handed it to her, saying "Here. Use the bag."

Fortunately, she was able to get it open just in time. She spent most of the next 5 minutes hurling what smelled like rum along with whatever she had eaten that night. When I thought she was done I pulled a paper napkin out of my shirt pocket and handed it to her in case she needed to wipe her hands and face a bit.

"Wow," she said, "you are really prepared." I was glad she was coherent enough to notice. That was a good sign.

I could see her bag was nearly full, so I stopped at a scenic viewpoint near La Selva. Just to make sure, I gave her a clean bag as I took her full bag. I got out, tied it off, and dumped it in a nearby trash can. It was a good thing I gave her the second bag. She wasn't done yet.

She apologized several times. I told her not to worry. I just wanted to make sure she was going to be okay. I told her she got big points for using the bags.

We eventually got to her home in Watsonville. She was able to thank me again and get out of the car unassisted. She even had the presence of mind to take the second bag with her (with the napkin in it, I guess) when she got out. She was walking okay as she made her way to her front door.

I spent a few minutes cleaning up. I'd gotten some liquid from the outside of the first bag on my hands as I was taking it to the trash can. So I pulled some alcohol wipes from the glove compartment and cleaned up my hands and the steering wheel.

I shut off the app and kept it off until I got back to Capitola, not wanting to be driving around in the boonies around Watsonville in the dark. I did one more trip, then called it a night at one fifteen. With many of the bars closing, demand was up. There was a surge on in Santa Cruz, but I didn't care. It was time to get some rest.

Chapter 9

Trips to the ER, Part 1

Probably a Good Thing I Didn't Turn Off the App

Sometime after midnight, I picked up 3 guys at JJ's Cantina in Soquel and dropped them at The Esplanade. At that point I had done 23 trips. I was tired and my back was sore, so I decided it was a good time to head home.

Something, maybe my intuition, told me not to turn off the app yet. Less than a mile from my exit on Highway 1 I got a request from a house not far from mine. The address looked familiar, one I had been to before, on a corner in the Carbonera neighborhood north of my little street.

The house was the one I thought it was. The app said the passenger's name was Sandy. She and her husband were waiting for me at the top of the driveway. It was clear something was wrong with her. I didn't recognize her as the woman I had taken home before, but as I've mentioned, I often don't get a good look at my passengers.

It turned out she *was* the woman who rode with me previously. At first I thought she was drunk, but it also looked like she might be in some kind of pain. As we drove away, I mentioned that I couldn't tell if she was sick or in pain. She replied it was both. I reached for a plastic bag. She said she already had one.

She described an internal problem she had been having, essentially her intestines not working. The doctors she had been seeing didn't know what was wrong with her, so they had just been treating the symptoms. I looked at the destination address she had entered. Then I looked at the map. I realized we were going to the hospital. When I asked, she confirmed it. We were headed for the ER. She had contacted her internist when the symptoms got really bad. He told her he would meet her at the hospital. I shut off any new requests from Uber so I wouldn't feel rushed once we arrived at the hospital.

I got her there and made sure she made it through the door, then I called it a night. I hoped she was okay. I found it difficult to drop someone at the ER, not knowing if they would be okay—not knowing If I'd ever find out what happened to them, or if I'd ever see them again.

Chapter 10

Cash Customers

Good Service has its Rewards

It started on a chilly Monday night in early December. I had just dropped someone off on Pacific Avenue in Santa Cruz. I guess I was on Highway 1 going east, maybe headed toward 17th and Brommer or maybe Capitola, when I got a request for a pick-up in the Watsonville area.

Once I accepted the trip, the requester, let's call him Mark, called and said I was picking up his girlfriend and taking her to the Four Seasons Hotel in Palo Alto. He gave me directions to her house on a little street near the Watsonville Airport. He said his girlfriend's name was Angela.

So, I picked up Angela. She was Hispanic. She brought a small suitcase. She looked small, though bundled up in a large coat, it was hard to know for sure. She wasn't much for conversation either. However, when we arrived at the swank Four Seasons in Palo Alto, she was able to give me very specific directions to the front door (it's a large complex). Clearly Angela had been there before.

A couple of months passed. Then on a Monday evening in February I got a similar request for a pick-up in Watsonville. Once again, Mark called, telling me I was picking up his girlfriend.

Angela got in. The Uber app showed the destination as The Hotel Valencia, a very nice hotel in the middle of Santana Row, San Jose's upscale shopping and dining area.

Again, Angela didn't say much, but she knew right where the hotel was. On a hunch, I gave her my card as we arrived, telling her she could contact me directly if she ever needed to make sure someone would pick her up. I didn't really expect to hear from her, but I've had a few clients who live in somewhat remote areas and appreciate being able to book a ride in advance. I give them the choice of running the ride through the app or paying me in cash.

For a while I thought perhaps Angela was a high-priced "escort." That turned out to be not the case. She describes Mark as her boyfriend. He lives on the East Coast, but travels quite often for business. They meet periodically at a nice hotel or resort for a few days. I've taken her to other Bay Area locations to rendezvous with Mark. I even met Mark, twice taking him to the airport at the end of a week he spent with Angela in Santa Cruz.

In addition to Angela, I have several other private clients, many of whom prefer to pay in cash. Part of this can be explained by geography, part by economics, and part by a legal system that is full of more contradictions than evangelical Trump voters.

There are easily as many people living up in the forested hills outside of Santa Cruz as there are in the city itself. Many of these neighborhoods take a while to get to, and some rural addresses can be difficult to find, especially for the novice driver, or someone who comes from out of town to drive in Santa Cruz (imagine a driver who picked up passengers at the San Jose Airport, brought them to Santa Cruz, and then decided to accept a few rides in the Santa Cruz area before heading back over the hill. Very quickly drivers learn the limitations of a navigation system in remote, mountainous areas where there is poor cell phone coverage, and not much in the way of streetlights).

There are also drivers who, given the option, will turn down (or accept and then cancel once the actual pick-up location is revealed on the driver app) a passenger who is far enough away that the trip is just not worth the time for the driver. Even after Uber added a charge for a longer than usual pick-up (most of which was paid to the driver), many drivers might still reject those trips, especially when trying to reach a certain number of rides to get a Quest bonus. Many drivers continued to do this, even as the amount paid out in the bonuses got smaller and smaller.

I generally don't turn down those requests, though there are some exceptions. I don't like the idea of leaving someone stranded. One Saturday night, after midnight, I had just dropped someone off at some apartments near the western edge of Santa Cruz. I responded to a request from someone in Davenport, a tiny town along Highway 1 as it goes west and north from Santa Cruz toward Half Moon Bay. It's less than 10 miles away and takes about 15 minutes to get there.

As soon as I accepted the request, I got a phone call from the young woman who requested the ride. She begged me not to cancel, saying three other drivers had already done so. I assured her I'd be there soon. Apparently she and her friend were on a long day trip with a guy who wasn't exactly her boyfriend, but there was some kind of relationship potential, I guess. They were headed south from Half Moon Bay and stopped in Davenport for a beer. There he ran into some guys he knew and decided to stay and drink with them for a while. As the night wore on, my passenger Leah and her friend Michelle had simply tired of waiting and decided to Uber back home.

It turned out to be a long trip, as they were going over the hill to Saratoga. It was a $40 fare and she tipped me an extra $10. It was a good way to end the night.

Passengers who live in some of these remote locations tell me they can't ever be sure if someone will come there to pick them up, especially late at night or very early in the morning. When they say that, I hand them my business card and say, "Call or text in advance. Tell me where you're going, and where and what time to pick you up. If I'm available, I'll be there." You can imagine how reassuring this is for someone who must catch an early flight at the airport, or who knows in advance they're going to be drinking and will need a ride home from the bar or a party late at night.

This has provided significant additional income as that part of my business has grown over the past few years. Many of my current cash customers were referred to me by one of their friends who is also a cash customer.

I'll acknowledge that this side of the business requires some flexibility on my part. There have been times when I was already in bed and heard my phone vibrate (ringer off). On the other end is one of my "regulars." It's late and he needs a ride home. I accommodate these requests whenever I can, even if it means getting out of bed to do it. It's probably best that I don't really have much of a social life, thus I've never been in bed *with someone else* when those calls have come in.

The cash aspect is a factor of two things. I give these private passengers the option of running the trip through the Uber of Lyft app, or just paying me directly. For some, using the app is more convenient. But there are others, especially those who work in the cannabis industry (and dancers who make most of their money in tips), for whom the cash economy is simply the way business is done.

Even though cannabis has been legalized at the state level, it is still on the books as a series of federal crimes. As a result, banks are not allowed to provide business accounts for any cannabis-related operations. These include growing, testing, extracting, transporting, and selling, either wholesale or retail, cannabis products.

In the Santa Cruz area, that means a significant portion of the workforce is getting paid in cash. They don't deposit that cash into a bank account. That could produce some untoward attention from the Internal Revenue Service. So, people who are paid in cash pay their bills in cash. They also like to pay their driver in cash.

There's an old joke I heard years ago from a guy in Mississippi: What's the definition of a successful farmer? That's a guy standing in a field who has a wife with a job in town.

That's sort of how it works with people in the cannabis business, too. The spouse has a job. That provides some cover for how the rent and utilities are paid. Since the spouse has a bank account, he or she likely has a credit and/or debit card, too. That lets the spouse have an Uber app (because payment is always linked to a credit/debit card or a PayPal account).

So I might initially meet a person when the spouse requested the ride using the app. Perhaps I pick them up in Santa Cruz where they've gone out to dinner or a concert. If they live up in the hills, I offer a card. The one who works in the business asks if they can pay in cash if they contact me. I always give them that option. For me, that's part of giving good service. And to be honest, there's something very appealing about getting paid in 50 and 100 dollar bills.

Walter

My relationship with Walter began, as I recall, on a Saturday night in the fall of 2016. I picked up Walter and his wife Mia at a popular restaurant in Felton, and took them a little further up Highway 9 as it winds through the San Lorenzo River Valley, to a bar in Ben Lomond. In the course of our conversation, Walter mentioned they lived in Boulder Creek (another community still further up the Valley), and sometimes had difficulty getting a driver to come up there for a pick-up. So, I gave Walter my business card, and told him I'd be happy to make sure he has a ride if he can coordinate with me in advance.

Since then, Walter has contacted me many times. While he doesn't often give me a lot of lead time, I've found I can be very flexible for clients who pay me with hundred dollar bills.

Walter is smart, an astute businessman, and he likes to party. He is also a player in the San Lorenzo Valley's thriving cannabis industry.

He originally began using my services to attend sports events. He gets season tickets to a particular Bay Area pro sports team each season. He doesn't attend all the home games. Instead, he gets a "slice" of the home games, so every couple of weeks during the season, I'd pick up Walter, sometimes with his wife, other times with friends or business associates, and drive over the hill to the sports arena where his team plays.

Walter likes to have a few drinks before and during the games, so it wouldn't work for him to drive to the games himself. He generally doesn't have any trouble getting an Uber back to Boulder Creek after the game, although Bay Area drivers sometimes get a little freaked out driving Bear Creek Road, a winding, narrow road through the Santa Cruz Mountains, at night. But it is the shortest route to Boulder Creek from Highway 17.

After taking Walter to a couple of games, I started getting other requests from him. Sometimes it was just to get him home from a bar late at night. Other times I'd take him to the airport on his way to a business convention.

Walter has referred several of his friends and business associates to me. I get the impression he has given them instructions to "take good care of me." A couple of times Walter has even checked with me to make sure I was "being taken care of." In Walter's mind, that means a hundred dollars (one way) for a trip over the hill (San Jose and beyond), a minimum of fifty for a shorter trip. About once a month Walter attends a poker game with a few of his buddies. That often generates a ride to the game for one or more of the players. Occasionally, one will contact me after the game for a ride home.

These are not your average neighborhood poker games. In addition to there being plenty of beverages and weed on hand, they are catered by a very nice woman whose day (actually night) job is tending bar at one of the San Lorenzo Valley's popular watering holes. She brings an assistant, and the two of them make sure the food is excellent and the service is top notch. They stay until the game is over, often in the wee hours of the morning. I know this because they have been there finishing up when I've arrived, having been contacted to give one or more of the guys a ride home.

Walter and his friends all know they are paying more than they would if they were running the trip through the Uber app (if they were even inclined to use it), but they like the idea of having their own private driver, someone they can trust to show up to pick them up and get them where they need to go, regardless of the time of day or night. They've also learned they can trust me to be discreet about their personal and professional activities. They value that kind of service much more than the convenience of an app on their phones.

It's a relationship that has been mutually beneficial. During the slow months, which in Santa Cruz is generally from New Years Day until around mid-May, those cash customers have kept me afloat financially. I don't know if I would have survived as a rideshare driver without the additional cash flow they provided me.

Bad Night on Zayante Road

During the winter of 2018-2019, the Central Coast of California got a lot of rain. I mean *a lot* of rain. Lots of rain can make Highway 17 more dangerous than usual. When the pavement is wet, it gets slick. Drivers who don't slow down for those slick curves often end up in a bad way when their car or truck spins out of control. Sometimes they hit other vehicles. Sometimes they hit the wall and end up facing the wrong direction as other cars are coming around the same curve. If they're lucky, nobody hits them.

Rain also causes rockfall and mudslides, like the big one in 2017. By the winter of 2019, much of the damage to secondary roads in rural areas still hadn't been repaired from the floods of 2017. There were dozens of places where roads were squeezed down to a single lane. In those spots, stop signs have been placed on each end of the single lane portion—one direction has to stop while the other direction goes.

On Valentine's Day, things really turned to shit. That morning Highway 17 was shut down due to a large tree falling, hitting a couple of vehicles, and blocking traffic in both directions south of the Summit.

Walter had contacted me a couple of days before, asking if I could take him, his wife, and his parents over the hill to a game on the 14th. My schedule was open, so we booked it for a 5:15 p.m. pick-up.

It rained hard that day. I had seen some advisories about problems on the 17. Around 4:30 I got a text from Walter, suggesting we leave earlier due to the weather. I got to his house a little before 5.

By then, 17 was still closed, and Bear Creek Road was also closed, with both lanes blocked due to mudslides and rockfall.

Going north on Highway 9 might have been an option, but downed trees and power lines had shut it down, as well.

That left only one practical option, and that was Zayante Road. Zayante Road begins in Felton, the first town north of Santa Cruz in the San Lorenzo Valley. From Felton it goes generally northeast following Zayante Creek as it winds through the communities of Olympia and Zayante, then turning north through the mountains and over the summit, passing several mountain vineyards, until eventually intersecting Highway 35 (Summit Road) less than a mile from where 35 crosses Bear Creek Road, which was open from there to Highway 17.

While Zayante is a bit more narrow and curvy than Bear Creek Road, it was still a decent option. I had driven it coming back from San Jose to Scotts Valley just the night before, so I knew it could work.

With Walter in the front seat, and his wife Mia and his parents sharing the back seat, we left the community of Boulder Creek going south on Highway 9, passing through Brookdale and Ben Lomond. Once in Felton we caught Zayante Road north and east up into the hills while a steady rain was coming down.

It's about 11.4 miles from Felton to the point where Upper Zayante Road intersects Summit Road (Highway 35). The first few miles are relatively flat and straight. But Zayante Road follows Zayante Creek, and the road twists and turns as it goes further up into the hills, gaining nearly 1,700 feet of elevation. The road crosses the creek several times, with numerous narrow switchbacks.

Since it was winter, darkness came early. My car was at the front of a line of 5 cars. I could see their lights in my rearview mirror. We began to encounter cars coming the other direction. Lots of cars. And then more cars. The Silicon Valley commuters had figured out Zayante Road was their only route back to Santa Cruz.

While it was a little tight with so many cars going in the opposite direction on such a narrow mountain road, I could manage the driving, despite the rain and the darkness. That was until we got to a spot where a rockslide shut the road down to a single lane. We had a problem. Cars from the other direction continued past the slide without pausing, giving us no chance to get beyond it.

After sitting for a minute, hoping for a break in the steady flow of southbound traffic, Walter decided it was time to take some action. Lamenting that he was wearing his good shoes, he opened the passenger seat door and stepped out into what surely was some very wet mud.

Between the headlights and the darkness, I had a hard time seeing him. I think, between cars, he had moved past the rockfall, and attempted to find a driver on the other side who would be willing to let us pass. What I heard didn't sound good.

Talking with the first driver on the other side of the rockfall Walter said, "California law says cars going uphill have the right-of-way."

The other driver responded, "There are at least 200 cars behind me trying to get home tonight, so you can take your right-of-way and shove it up your ass."

Walter is an astute businessman, and a good negotiator. There were probably several bad outcomes that could have followed that interchange, but instead Walter chose to just walk up to the next car in the line to make our case.

It took a couple more tries, but he eventually convinced a southbound driver that the fair thing to do, since there were only a handful of northbound cars trying to get past the rockslide, was to do a one-for-one pass in each direction. Then he quickly ran back past my car to let the drivers behind us know what was happening. When he told me, he said, "Try to find a safe place to pull off about 50 feet past the rockslide and wait for me there."

Then he moved close to the rockslide, judiciously directing cars until all five from our side had passed. There really was no good place to pull off, so once I was far enough past the slide that all the cars behind me could clear it, I just waited with my flashers on until Walter returned and got back into the car. We made it about a half mile until another slide pinched the road down to a single lane again.

By now Walter knew the drill. His shoes were already trashed, so what the hell? He jumped out and walked down the line of oncoming cars until he found a reasonable driver willing to let alternating cars go past. It didn't take long, and we were on our way up a winding road of oncoming headlights in the rain.

We eventually got onto Summit Road. Walter and I consulted and decided that even though Summit Road goes south to get to Highway 17, it was better than taking a chance continuing north to Bear Creek Road which goes mostly north to 17. We'd be backtracking a little, but would probably make better time on the better road.

I got them to the game just before it began. I heard on the radio that crews from Santa Cruz County had shut down Zayante Road not long after we had made it through. Highway 17 southbound had reopened by the time I brought them back over the hill after the game.

A story the next day in the San Jose Mercury News mentioned the closure on Zayante Road that night. The report contained the first use of the term "atmospheric river" I had ever seen. Thanks to Climate Change, it would become a common weather descriptor over the next few years.

The After the Game Party Place

While Walter often brings his wife with him to watch a pro sports game (she's at least as big a fan as he is), other times I would take Walter and one or two of his buddies. On one of those nights, Walter heard about a place where there was a big private party after the game.

At that point in our relationship, I wasn't picking them up after the games yet. That night, after dropping off Walter and his business partner Vinny at the game, I drove back over the hill to Santa Cruz and did a regular night of driving. I got to bed about 1:00 a.m. I guess I wasn't in deep sleep when I heard my cell phone vibrate.

I looked at my phone. It was a text from Walter, wanting to know if I could pick them up from a party. He gave me the address. I told him I'd be there in about 40 minutes. I got dressed quickly, got in my Subaru, and headed north on Highway 17.

The address was an warehouse in a new business district in Silicon Valley not far off the US 101 Freeway. I don't know what goes on at those parties. All I know for sure is Walter and his buddies are pretty drunk by the time I pick them up, usually between 1 and 3 in the morning. I'm actually impressed Walter was able to text me the address that first night.

When I got to the party location, it was almost 2:00 a.m. There were other Uber and Lyft cars making pick-ups there, too. A guy who was doing security walked up to my driver's side window and asked who I was there for. I gave him Walter's name. He walked inside, was there for less than a minute, then came back and told me they would be right out.

Soon Walter and Vinny came out, got in the back seat, and we headed back. I think they were both passed out before we got to Highway 17 heading south. I woke them up on Bear Creek Road as we were approaching Highway 9. Walter was surprised he had slept the whole way. He gave me a hundred dollar bill for getting him home.

I took Vinny to his apartment in Santa Cruz. He gave me a fifty. It was a long night, but financially rewarding. The next time I saw Walter he told me how much he appreciated me getting him and Vinny home, especially given how drunk he was. Then he checked to make sure Vinny had "taken care of me," too.

From then on, whenever I took Walter to a game, I arranged to be available to pick him up, no matter the time. I did that post-game trip several more times. Sometimes I would pick up Walter and Mia at a pre-arranged location near the game and take them home, other times I'd pick up Walter and one or two other guys and take them to the party location. The last time, Walter had an old friend with him who was visiting from out of town. Perhaps Walter didn't have tickets for the game that night. I picked them up in Boulder Creek around nine o'clock. I drove them straight to the party location. Before getting out of the car, Walter gave me 2 hundred dollar bills. I assumed he was paying me in advance for the return trip.

Around two in the morning I hadn't heard from Walter, but I got up and started driving. I hadn't been on the road for 5 minutes when I got Walter's text.

I picked them up at the same place. Again, they both slept the whole way back over the hill. When we got to Walter's, he gave me another hundred dollar bill, and asked me to drop his friend off in Ben Lomond.

I'll just note that the only times I've made $300 a night as an Uber driver, it has been on Halloween or New Years Eve, the two busiest nights of the year for an Uber driver in Santa Cruz.

From then on, game nights when Walter was with friends, as opposed to family, often became a 3-legged affair. I'd take them to the game, keep track of the game on the radio and when it was into the final period, I would head back to the sports venue to pick them up and take them to the post-game party location. Sometime around 2:00 a.m., I would drive back toward the party location, and wait for Walter's text asking me to pick them up for the trip home. I would get paid for each leg.

One night I picked up Walter and Stephen, one of Walter's business associates, around 2:30 a.m. There was a delay. Apparently Vinnie wasn't ready to leave yet. After waiting a few minutes, Walter made the command decision, and we left, leaving Vinnie to party on and fend for himself.

Later that morning, around six o'clock, I got a call from Vinnie, wanting to know if I could come back over the hill to pick him up. At that point I was making breakfast and getting ready to take my daughters to school. I felt bad, but I had to turn Vinnie down. A man's got to have his priorities, after all.

All good things come to an end, I guess. One night, after a game, I picked up Walter and Stephen and drove them to the warehouse where the previous parties had been. There was no activity outside. The door was locked and there was a For Lease sign on the door.

But Walter is a very resourceful individual. By the next season he had found another late night party place.

Chapter 11

Lost & Found, Part 2

Tracking Down the Keys

Sometimes I've been known to go above and beyond what most rideshare drivers would do. I'm sure there are one or more reasons why I do that. I don't know if I can articulate them, except to say that I try to do what's right when I can. That plays out most often when it involves an item left in my car by a passenger.

On the evening of Valentine's Day, 2017, I picked up two female students at Stevenson College on the UCSC campus and took them to a bar downtown.

Not long after, I picked up four more female students at College Nine on the UCSC campus and took them to The Crepe Place on Soquel. During that trip, one of the women in the backseat discovered a set of keys on the floorboard. The woman sitting next to her said, "Oh, those must have fallen out of my bag," took them from her friend, and dropped them in her bag. She was memorable, as she had a distinctly Australian accent. Her friend called her Laura.

Later that night I received a message through the app from Susan, one of the two women I took downtown, saying she had lost her keys and was wondering if I'd found any in my car. I replied that I hadn't found any keys, but I'd get back to her if they turned up.

The next morning, as I recalled the events of the previous night, I realized it was possible the keys Laura claimed were hers might have been Susan's. That got my investigative juices flowing.

I didn't know Laura's last name, but I did know I picked up her and her friends up at College Nine. So I got online and discovered College Nine was the home of UCSC's International Living Center (ILC), a community of 150 students from around the world. Conceivably, a student from Australia might be part of that community.

Next, I tracked down the names of a couple of the Resident Advisors at the ILC, and sent each of them an email, explaining who I was and what I thought may have transpired the previous evening, hoping one of them might be able to assist in finding Susan's keys.

I heard back from one of them, an RA named Will McAfee. Here's the email conversation:

> Dear Mr. McAfee, I could use your assistance.
>
> I'm an Uber/Lyft driver. Last night (Feb. 14) one of my passengers lost her keys. She thinks she might have left them in my car.
>
> On a subsequent ride with 4 students I picked up near Building 4 of the College Nine Apartments, one of those students found some keys on the floor or seat of the back seat of my car. One of the women in that party thought they were hers and that they had fallen out of her purse. She accepted them from her friend and placed them in her purse. I dropped this group off at The Crepe Place.
>
> The woman who claimed the keys is named Laura. She is from Australia.

If possible, I would like to check with her to see if, perhaps, the keys were not hers. I would then be able to facilitate the return of the keys to the student who lost them.

I would appreciate any assistance you can provide in this matter.

You can reply to this email, call me, or text me. My cell phone number is 831-555-XXXX.

All the best,

Burt Hays

****

Greetings Burt

I've been in contact with Laura, and she has a set of keys which aren't hers.

She or I are willing to meet with you to return the keys to you at your earliest convenience

You can email me, or call me on 650-555-XXXX

Best,
Will

William McAfee
University of California, Santa Cruz

Residential Assistant, College Nine
International Living Centre

****

Many thanks, Will. The student who lost them is free until 1pm. I'm about 20 minutes from campus. How are the schedules for the 2 of you?

Burt

****

I'm free until 2pm, and Laura is heading back now to give me the keys. However, if we can't meet up I can leave the keys at one of the housing offices around campus and the student can pick them up there.

Will

****

I'm meeting the other student at Stevenson Circle at 12:30. Shall I connect with you to pick them up before that, or would you &/or Laura like to meet us there?

Burt

****

I can meet you there, that works for me.

Thanks,

Will

****

Great. See you then. I'm driving a blue Subaru Impreza. I've got a fairly long beard.

Burt

****

Great, thank you

****

The handoff went perfectly. Will exercised an appropriate amount of discretion. Laura and Susan never met face to face, so there were no opportunities for accusation or acrimony. Most importantly, Susan got her keys back. Mission accomplished.

Fell Out of the Bag

Not very often, maybe once every week or so, I'd get a request for a pick-up at a grocery store. These are not high dollar trips. They are usually short, often two miles or less. Loading and unloading the grocery bags can take up nearly as much time as the trip from the store to the passenger's residence.

But I don't mind giving these rides, mostly because quite often the passenger doesn't have other options. If they don't have a car and they have to buy food, what else are they supposed to do? I guess before rideshare drivers came along, they would have to call a taxi.

One Sunday evening I got a request for a pick-up at the Lucky market in Capitola. The passenger was a guy probably in his fifties. He was waiting for me at the curb in front of the store with three shopping bags. I opened the hatchback so he could put the bags in the space behind the back seat.

I drove to his home, about a mile away in the Pleasure Point neighborhood, and pulled into his driveway. He said, "No need for you to get out. Just pop the hatch. I can get the bags."

It was around 10 when I called it a night. I did my usual inspection of the car. There in the storage space behind the back seat were two bottles of extra virgin olive oil. They must have fallen out of one of his shopping bags.

I drove back to the house where I had dropped him off. The house was dark, and by then it was around 10:30, so I didn't ring the doorbell. I knocked softly, just in case someone was still up. Getting no response, I left the bottles of olive oil there by the carport door.

A few days later I was scrolling through my ratings and feedback from passengers. There was a comment.

> You gave me a ride from the store on Sunday night. When I got home and started putting away the food I bought, I was missing a couple bottles of olive oil. They must have fallen out of one of my bags and I didn't notice it. I thought, *Well, I guess I'll never see those bottles again.* Then when I stepped out the door the next morning, there were the bottles, right by the door. Thank you! Excellent service. Definitely 5 Stars.

I don't think Uber actually reads passenger comments like that, except maybe the 1 Star comments. All they do is count the stars and crunch the numbers. Even so, it was nice feedback, and it made me smile.

Chapter 12

Hurlers, Part 2

Round & Round

I frequently do pick-ups and drop-offs at the Chaminade. It's a nice, upscale resort, on 300 acres of forested hillside land on the north edge of Santa Cruz, that began as the summer lodge of a prominent Santa Cruz attorney in the early 20th century. Following his passing in the 1920s, it became a religious retreat for the Brothers of the Society of Mary.

In 1930, the Brothers opened Chaminade High School on the property, named after William Joseph Chaminade, a French priest who founded the Society of Mary following the French Revolution. The school was closed in 1940 in the midst of difficult financial times. The Brothers then leased the property to the Mary Burke Foundation, which ran a school for children with developmental disabilities until 1949, when it closed due to financial problems.

The Brothers returned to Chaminade and founded The Postulate, a school for young men who intended to join a religious order. The school was eventually moved to Los Angeles, and the Brothers put the property up for sale.

It was purchased by three partners who, in 1985, opened the Chaminade Resort, keeping the name of the old high school.

In addition to being a nice, secluded place to stay when visiting Santa Cruz, the Chaminade is also a favorite spot for weddings, receptions, and other large social events.

It was at one of these events, a wedding and reception, as I recall, where I arrived one Saturday evening. There had been a big, catered Italian meal, followed by a live band. There was a lot of dancing.

A dad and his daughter, I'd guess she was about 10-years old, were my passengers. They both got into the back seat. Apparently the daughter had been dancing, spinning round and round in circles for some time, prior to my arrival. No sooner had she sat down in my car than all that spinning caught up with her. Suddenly, there was, or what must have previously been, lasagna, wedding cake, and punch all over my back seat, as well as all over her father's nice clothes.

Standard procedure when a passenger makes a mess is for the driver to take photos and attach them as evidence when filing a report with Uber. But this was not a standard situation. Nobody was drunk, so I wasn't handing out barf bags prior to the incident.

The father was very apologetic. He opened his wallet and asked me if a hundred dollars would take care of it. I made a split-second decision, accepted the cash, didn't file a report, and drove off to find a place to clean up the back seat.

I had made the decision a few months before to spend money on some faux leather seat covers. At that moment I was very glad I did. It made the clean up much easier than if I still had cloth-covered seats.

Worse Than I Thought

In 2018 the 4th of July fell on a Wednesday. That last weekend of June was unusually slow for a weekend in the summer in Santa Cruz. I can't say why for sure. With the 4th occurring in the middle of the week, maybe people were so confused about which weekend should be considered the 4th of July weekend, they took a pass on the front end, saving their energy, their money, and their livers for a long weekend on the back end.

Still, that last weekend in June was not uneventful. There are always lots of weddings in June in Santa Cruz, and that last weekend was no exception.

Since business was a bit slow that Saturday night, I stopped at my house for a little break around 10:00 p.m. That's a time of night when the ride business tends to drop off a bit. It seems to be when people have gotten wherever they are going, but it's not time to leave yet.

After using the bathroom and stretching a bit, trying to work some ache out of my back, I got in my reclining chair, hoping to catch a short nap. I don't think I actually fell asleep, but at least my back got to spend a little time in a position other than sitting in my driver's seat.

I was back on the street before eleven. Since my neighborhood isn't far from where Ocean Street becomes Highway 17, I got a request for a pick-up in Scotts Valley, which is just 3 miles north.

The requestor's name was Danielle. The pick-up location was a spot on Disc Drive where there often are holiday parties. That night, there was a wedding reception.

I had to wait for a few minutes with my flashers on before Danielle came out to find me. She had arranged a ride for some other people, but they weren't there yet. Herding drunks is worse than herding cats.

Eventually, two others showed up. One was a guy Danielle had to talk to for a bit before bidding him farewell. His name was Trip. The other was a woman who might have been Trip's mother, maybe Danielle's mother, maybe both. I wasn't that interested, so I didn't pursue their familial connections.

It was clear from the get-go that Trip was drunk. Danielle and Mom had to help him get into the car and get his seatbelt fastened. I'm guessing they had been drinking, too (it was a wedding reception, after all), but probably not as much as Trip.

Once they were both in the back seat, I asked Trip if he was okay. He assured me he was. I asked if there was any chance he was going to get sick. He assured me he wouldn't. I told him there were bags right in front of him, in the pocket at the back of my seat. He seemed impressed that I was so prepared.

Their destination was Corralitos, a very rural farming area north of Watsonville near the southern end of the Santa Cruz Mountains. It's about 20 miles from Scotts Valley. The route takes Highway 17 back to Santa Cruz, picks up southbound Highway 1 toward Watsonville, then heads out into the hinterlands at Freedom Boulevard. It was late on a Saturday night, so I was focused on the traffic on the highway, not so much on my passengers.

Trip had his window down as we got on Highway 17. I usually have windows up at freeway speeds just to reduce the noise, but I'm okay if a drunk wants to have his window down. Fresh air is usually a good idea in those situations. By the time we got on to Highway 1, I noticed Trip was spitting out the window every minute or so. I thought maybe he had some kind of mucus drainage going on, so I pulled a pack of facial tissues out of the console and handed them to Trip. Nonetheless, the spitting continued.

After passing Aptos High School, Freedom Boulevard becomes a rural, winding road passing first through large-lot residential areas which eventually blend into small farm country. It's dark out there late at night.

Corralitos isn't really a town, more of an unincorporated community of around 2,500 people. Housing gets denser toward the center, though you wouldn't know that in the middle of the night. You know you're there by the handful of commercial buildings at the 4-way stop.

I think it was at that intersection, with some headlights coming up the same road a few hundred yards behind me, that I noticed what looked like something splattered on the rear window of my Subaru hatchback. I began to wonder if some of Trip's spitting had somehow found its way to my rear window.

After a few turns down what are best described as paved farm roads, the GPS system found our destination, a large house about 50 yards off the road. My passengers got out. At least Trip was able to walk to the house. I got out to inspect that splattering on my rear window. What I found wasn't pretty.

Trip had apparently been throwing up, catching it in his mouth, then spitting it out the open window.

Or trying to. There was puke all over the inside of the door and window, more outside on the side of the car, and on the rear window. First I took photos with my phone, then I commenced to clean up as well as I could, using a bottle of water and a pack of tissues. It took a while. About 20 minutes into the project Mom came out to see why my car was still there.

She said, "Oh did he throw up? I didn't realize he'd done that. I'm so sorry." I continued to wipe up as much as I could, hoping Trip had consumed enough alcohol to kill whatever little infectious critters might be lurking in all that puke.

Eventually I'd done what I could with the supplies at hand. I left a pile of puke-covered tissues on the ground. I didn't want them in the car.

There is no cell service out there in the boonics, much less a data signal. So neither Google Maps nor Waze was any help for finding my way back to the highway. I did my best to navigate in the darkness, only making one wrong turn before realizing I was headed toward the mountains, not the highway.

I eventually found a road name I recognized. I'd been on it a few times before and I knew it would intersect Freedom Boulevard which would take me back to Highway 1. I drove to a self-service car wash in Capitola and parked in a bay to spray off the residual puke. The token machine didn't take dollar bills. All I had were three dollar bills and a twenty. I didn't want 20 bucks worth of tokens.

Since I had phone service again, I filed a puke report with Uber, including the photos I had taken prior to the rudimentary cleanup. Then I drove to another car wash that I knew would accept dollar bills. There I sprayed the car off as well as I could. With some puke still evident on the fabric of the door panel, I knew I was done for the night. So I went home and got some sleep.

I awoke in the morning to a message from Uber that the customer had been charged $150 to cover my cleaning costs. I was pretty sure I could do it myself for considerably less, but the extra money would at least help cover some of the money I wasn't making until the car was thoroughly cleansed of the vomit.

Inspecting my car in the light of day, I noticed some residual puke in the interior I had missed the night before. The dude had actually managed to spit vomit on his own backseat headrest. That's not easy to do. I hope he washed his hair when he got home.

Chapter 13

I'm With the Band

Skip the Line

The Catalyst is the main music venue in downtown Santa Cruz. There are two floors. One showcases the main act, the other has the lesser known up-and-comers.

I must confess I'm not knowledgeable about many of the touring musicians of the current millennium. But the Catalyst books a lot of popular performers, judging by the lines of people that stretch along Pacific Avenue and sometimes around the corners in each direction—one line to buy tickets, the other for people with online tickets waiting to get in when the doors open.

Fans come from as far away as Fresno and Sacramento to see their favorite bands perform. They'll book hotel rooms, spend time at the beach, hit the Boardwalk, and make a weekend mini-vacation out of it. As a woman from the Central Valley told me, "We come to Santa Cruz in the summer because it's cooler, and we come in the winter because it's warmer. And summer or winter, the music is always better than what we get at home."

One Saturday night during the summer, the band must have been very popular. It was the first time I had ever seen the line go south down Pacific Avenue to the corner, continue west on Elm Street for another block and turn again, heading north on Cedar Street.

I can't recall the name of the band that was headlining The Catalyst that Saturday night. Maybe it was The Twisted Pears, or maybe Hoosier Daddy (Straight outta Bloomington!), or maybe it was Piddy da Foo, or maybe it was Ivanka's Alibi. All I knew for sure was, whoever they were, they must be one hot band.

Around quarter til 9 I got a request for a pick-up at the Holiday Inn Express. There I picked up three young men, two white guys and one black guy. Their destination was an address on Pacific. We started heading downtown.

I checked the address and realized they were going to the Catalyst. Thinking about those lines I'd seen earlier wrapping around the block I jokingly said, "Well, unless you guys are the band, you're not going to be able to get in. I'm pretty sure there are no seats left."

The guy in the passenger seat said, "Is it that obvious?"

I said, "What do you mean?"

"Dude, we are the band."

He told me their band's name. Like I said, I can't remember. But they seemed like great guys. Apparently they play some hot Reggae, and a lot of people really dig it.

I think I should try to catch their show the next time they come to town.

Knee Deep in the Oldies

I'm reluctant to tell this story. As I said in the intro, I've changed people's names. But obviously certain details could blow the anonymity of an individual, especially if that individual is semi-famous to begin with. Like the lead singer in a rock band, for instance.

It was graduation weekend at UCSC. There were lots of families in town, many of them needing rides to campus as parking at UCSC was limited. I'd made a couple of trips to campus already that morning. I knew the drill.

I got a request for a pick-up at the Hilton Santa Cruz, which is actually in the town of Scotts Valley north of Santa Cruz, just off Highway 17.

There was a couple outside the main entrance waiting for me when I pulled up. They got in.

There's a brief routine I always go through before I start the app rolling. First I verify I've got the correct passengers (the ones the app says I'm supposed to pick up – learned that lesson once and never made that mistake again). Then I ask my passengers to fasten their seat belts. Both have to happen before I put the car in Drive.

So I said, "Good morning. Tell me your name please."

The man replied, "Michael, or maybe Mickey."

I checked the app. I was supposed to pick up someone named John.

I replied, “Hmm. The app has another name listed. Maybe I’m here to pick up someone else. Is there a chance Uber knows you by a different name?”

“Oh, yes, I suppose it could be John. That’s my first name, but I go by Michael or Mickey.”

“Okay, Mickey. John was the right answer. Please make sure your seatbelts are fastened. And we’re going to the university for graduation today?”

“Yes, my son is graduating.”

“Well, congratulations, then.” And with that, we were on our way to UCSC graduation.

Not long after making the transition from Highway 17 to Highway 1 and Mission Street, Mickey asked, “Do you think we’ll have any trouble getting a ride back to the hotel after the ceremony? I need to catch a flight to New Orleans that leaves at two.”

“Shouldn’t be too difficult, although there will be a lot of friends and families looking for rides at the same time. Here, let me give you my card. Call when you’re ready, and if I’m available, I’ll get there as quickly as I can.” Mickey took my card.

“What’s happening in New Orleans?” I asked.

“I’ve got to be there for a concert.”

“To attend, or are you performing?”

“Well, I’m performing.”

“Really? Vocal or instrumental?”

“Uh, vocal.”

At that point, Rachel, Mickey’s wife, joined the conversation. “Mickey, why do you always pussyfoot around like this? He’s the lead singer in the [?????].”

[Author's note: Okay, for the sake of playing the anonymity game, let's say the name of the band was The Enterprise. But it had been through a bunch of changes over the course of the 50 years leading up to that day when I gave Mickey and Rachel a ride. Maybe before it was the Enterprise it was known as the Hamilton Enterprise, and before that it was the Hamilton Avione. And Mickey sang with another band before he joined the Hamilton Enterprise. Okay, I'm glad we got all that out of the way.]

Now, in truth, I wasn't aware of all that at the time I was giving them a ride. It required a little research on my part, after the fact. I had been a fan of both the Hamilton Avione and the Hamilton Enterprise back in my reckless youth, but by the time Mickey joined the band I had joined the army, and I was paying more attention to the Russians than I was to pop music.

Obviously I was aware there had been some significant changes to the band's personnel over the decades. But I wasn't very current in my information. So I asked Mickey how it was going with his current set of bandmates.

I think he got a little defensive, maybe because The Enterprise was pretty far over the hill at that point. I mean, hell, Mickey is a few years older than me, and while I may not be over the hill yet, I've got a pretty decent view of the summit.

"I've got a great band," he said. "I'm very happy with them." By then he wasn't just the lead singer in the band. It was literally his band.

But he was curious about what I actually knew about his band. He asked me if I liked their music. Frankly, I could only name two of the songs from his era with the band, and they were pop hits 30 years before that graduation weekend. He asked me about one of the songs in particular. "Did you like it?" he said. "Do you think it's a good song?"

"Definitely a good song," I replied. It felt like he wanted more. But I really didn't know enough to give him more. "Yeah, a good song," I repeated. It was a good song, a beautiful love song, though frankly it wasn't a record I owned or ever put on a playlist. In a different year, it might have won a Grammy. But that year, in the mid-80's, it couldn't compete with hits by Don Henley, Tina Turner, and Dire Straits. That's life in the music business, I guess. And I imagine it's tough when your glory days are three decades behind you.

I dropped Mickey and Rachel as close as I could get to the site of their son's graduation ceremony. Then I made my way back toward town.

Mickey did call me about an hour later to see if I could get him to the airport, but by then I already had another passenger and was headed for SFO.

I think Mickey did make it to the concert on time because I've seen the video. And for the record, I don't think there are many pop vocalists who are pushing 70 and can still hit the notes the way he does.

Chapter 14

Road Rage Slime

Strange things happen on Highway 17. There was the mudslide that buried a vehicle. There was the tree that was blown down during a rainstorm, hitting a vehicle and shutting down 4 lanes of traffic for the better part of a day. There are the regular occurrences of cars going too fast through the downhill curves, and skidding out of control into the rock walls on the side of the road, sometimes bouncing back out into oncoming traffic.

There was the brand new Chevy Camaro that was in front of me as we were southbound and approaching the top of the hill. About a quarter mile before the summit, the engine exploded, filling the road with black smoke so thick all I could see was the Camaro's tail lights as it coasted to the pull-off space once its momentum carried it to the summit. I don't know what caused the explosion. The smoke had eased a bit by the time I cautiously passed, and it didn't look like the engine was on fire.

Then there was the road rage slime incident.

It was late afternoon on Monday, July 3, 2017, the day before Independence Day. I was hoping to run up my ride numbers to catch a little bonus money. There were lots of tourists in town, lots of families going to the Boardwalk.

I had returned home for a bathroom break. Our house was very close to Highway 1, and a back road through the neighborhood is a shortcut to the 17, so I frequently get requests for rides that are somewhere along the highway while I'm sitting in our driveway. That's what happened. The pick-up was in Scotts

Valley, just 5 minutes from the junction of Highway 17 and Highway 1.

The passenger was going to Ben Lomond, which is on Highway 9, a couple of miles west of Scotts Valley and north of Felton. Before I got him to his destination, I had a request for a follow-on pick-up. That one was also in Ben Lomond.

That passenger turned out to be Taylor, a young girl, maybe high school age. Her father had placed the request through the app, then sent me a message through the app. I was to bring her to their home in Saratoga, an upscale town over the hill in Silicon Valley. So I picked her up and we headed back through Scotts Valley to get on Highway 17 northbound.

I should point out that it is technically a violation to transport an unaccompanied minor. A bus driver can do it. I'm not sure if a taxi driver can. But according to the rideshare companies, we drivers shouldn't do it. There's just too much risk involved. My guess is it's a rule that was pushed through by lawyers for the insurance companies. That's a profession with infinite job security.

But if anyone was going to enforce that rule, drivers would be required to card every single passenger, making sure every passenger was either of legal age or was accompanied by an adult. And let's face it—nobody is doing that.

We were coming down the ramp, trying to merge on to 17. It was one of those awkward situations. There was a white van coming in the right lane I didn't see at first. I slowed down, staying to the side so he could pass. It would have been nice if the driver changed lanes in order to make some space for me. When he didn't pass or change lanes, I was running out of shoulder. So I ended up accelerating, trying to get out of his way by pulling away as I moved in front of him.

Maybe that pissed him off. Maybe he didn't like my ***America was Great when we had a Real President*** bumper sticker. Maybe it was something else further down the highway. Each of us changed lanes a couple times, speeding up, passing slower vehicles, then moving over to the right lane again. At one point I moved into the left lane and was in front of him again. I had signaled the lane change, and there was plenty of room. We were both passing a slower car.

Not long after that, both my passenger and I heard a loud thump, as though something hit the top of my car. I checked my rear-view mirror, but didn't see anything to explain it, like maybe a large bird falling from the sky then bouncing off. I thought it was curious.

The white van passed us on the right. With it ahead of us, I could see it was moving a bit erratically, having trouble staying in the center of the right lane. I decided it would be wiser to be in front of the van instead of behind. If he did lose control, I didn't want to be caught in whatever traffic mess ensued.

The left lane was clear ahead of me, so I picked up speed and passed the van. Not long after, traffic slowed in the left lane. As I was slowing down the right side of my car was suddenly splattered with a hefty amount of some sort of light-colored slime. It was all over the right-side windows, front and rear. The white van had passed us again, this time on the right.

I looked in the rearview mirror and said to my young passenger, "Well, Taylor, you are witness to the strangest thing that has ever happened to me as a driver."

"That is totally weird," she replied.

Again, the white van was swerving a bit. We crossed the summit and started into the downhill curves. I touched my phone, bringing up a keyboard. I called 911. It rang a few times with no answer. I needed to focus on the curves, so I hung up. I saw the white van pull off into an open space on the side of the highway.

Later it occurred to me I could have asked Taylor to try and get some video of the white van. Then I realized that would have been ill-advised. Her father expected me to get her home safely, not involve her in a road rage legal case.

As I was pulling up in front of their house, I hoped her father couldn't see the side of my car. If Taylor wanted to tell him about it, that was her call, but I didn't want to advertise what had happened to my car on Highway 17.

I located a nearby gas station with a drive-through car wash. It must have been a popular spot. I was the fourth car in line. The line was only moving forward every few minutes. That gave me a little time to get out and clean off the bigger pieces of what appeared to be something like whole wheat bread dough. I was glad it wasn't some industrial-strength chemical substance.

While I was waiting my turn, I got a call back from the 911 operator. Apparently, they follow up on calls that are not completed. I described what happened, and gave her a description of the van. I doubt the Highway Patrol found the van. It had been nearly half an hour since my car got slimed. I did my trip through the drive-through car wash. It made my Subaru look respectable again.

Emotionally, I felt pretty normal throughout the event, really just trying to get Taylor home safely in order to complete the mission. But once that was done, it hit me. I started feeling shaky, like my adrenals had overloaded and were blown out. I got something to eat, then decided to just drive back over the hill to Santa Cruz and call it a night. Since the next day was the 4th of July, it would likely be a full day of driving that could go late into the night, fireworks and all.

Chapter 15

Weed, Part 1

If You Feel Something is a Little Off

On December 22, 2017, Uber sent an email to drivers that included the following guidance:

> Starting January 1, marijuana will be legal for recreational use in California. This means adults over 21 can grow, purchase, possess, and consume specific amounts of marijuana under certain conditions.
>
> Just like drunk driving, getting behind the wheel while high on marijuana is illegal and can land you a DUI. ***Additionally, if you encounter a smelly or sleepy rider, or simply feel something is off after you end a trip, remember that 24/7 support is just a tap away.*** (emphasis mine)

I have occasionally remarked that guidance from Uber is often written by people who have never spent any time behind the wheel with the app on. The above message would indicate they have never driven in Santa Cruz, either.

Here's what I mean. On any given night, roughly half of my passengers might be stoned. So what?

Smelly? WTF? For me, cannabis has three distinct aromas. First, there's how it smells prior to burning. It is buds and leaves. I get wafts of it just driving down the street in town or on the highway going through the mountains. It's everywhere here.

One backseat passenger had an aroma of raw weed so strong, after I dropped him off, my next passenger asked me if I had just trimmed my plants (not something I do, but a common practice in Santa Cruz).

Second, there's the familiar aroma of weed being smoked. Again, you can almost get a buzz just walking down the street downtown. My next door neighbor fires up a little weed every afternoon when he gets home from work. They have a "no smoking in the house" rule, so he smokes outside. I smell it as soon as I step out the door or as it floats in through my bathroom window.

Third is the way it smells on people's clothing after they've been smoking. Since I don't smoke anything, perhaps my senses are keener than most. Weed smoke leaves a distinct aroma, very different from cigarette smoke. While I don't let anyone smoke anything in my car, lots of my passengers smell either like they just trimmed a bunch of plants, or they smoked a few bowls just before I arrived.

Santa Cruz's relationship with cannabis goes back a generation or two. This place was a bit of a counter-culture haven dating back to the late 1960s. Between the surfing culture, the university, and a vibrant arts and music scene, Santa Cruz has leaned toward the Boho side of life for a long time. Ironically, the War on Drugs solidified and deepened the town's relationship with weed.

During the 1980's (think Reagan administration) the DEA declared full-scale war against pot growers up north in California's so-called Emerald Triangle. Known as the Campaign Against Marijuana Planting (CAMP), it often involved agents in combat gear swooping into a growing area on helicopters, jumping out, and commencing slash and burn operations. The goal was to destroy the industry by destroying the genetic stock of cannabis in the region.

According to local cannabis lore, that's when a few visionary growers grabbed the seedlings from their favorite plants, packed up, and headed a couple hundred miles south into the hinterlands of the Santa Cruz Mountains. Here they continued cultivation, developing new and better genetic strains. They maintained a low public profile, the DEA never caught on, and pot cultivation became a silent feature in the local economy.

Not surprisingly, given the culture of Santa Cruz and the emerging cannabis industry, Santa Cruz was an important hub for the medical marijuana movement that swept California in the 1990s. By the time I arrived here in 2015, there were a half dozen medical marijuana dispensaries within the towns of Santa Cruz, Soquel, Capitola, and Aptos, and a few more in smaller communities up in the hills. In November of 2016, California voters approved a ballot measure legalizing cannabis for recreational use by adults. That took effect on January 1, 2018.

According to those who know such things, today Santa Cruz County produces some of the best cannabis in the world, and grows about 10 percent of California's supply.

So when Uber sends me an email about sleepy or smelly passengers while advising about our new cannabis law, I view it within the context of several decades of cannabis use in Santa Cruz.

Here's my perspective as a driver: While I've had people throw up in my car as the result of consuming too much alcohol, I've never had anyone hurl because they smoked too much weed. I'll take the aroma of cannabis over puke on my car seats any day. End of story.

Chapter 16

Weed, Part 2

Zeke the Outlaw

Of all my Cash Customers, Zeke was the one I liked the least.

I first met Zeke when his girlfriend requested a ride on the app. I picked them up somewhere in the hills north of Soquel, not far from a popular summer camp for kids.

At the time, Zeke didn't know I sometimes worked off the app, often giving rides to some of his associates in the cannabis industry. And I didn't realize he was in the business. On a subsequent trip, again one his girlfriend had requested, I brought him home. I realized it was the same house where I'd picked up Rocky a few times. Rocky was one of my cash customers and a good friend of Vinny who was Walter's business partner.

Realizing the connection, I must have given Zeke my card, suggesting he could contact me directly if he needed a ride.

While guys like Walter and Vinny worked strictly within the law (at least the state laws of California, once Medical Cannabis became legal), Zeke worked the other side. I think they all shared a warehouse in the East Bay where vaping cartridges were pre-loaded with a liquid cannabis product. Where their business models differed was in what happened to the product when it left the warehouse.

Walter and Vinny could see a growing market for the hybrid product in California, where weed was legal with a note from your doctor, then for everyone over the age of 18 starting in 2018. So they threw in with some other entrepreneurs renting space and leasing equipment to fill vaping cartridges. They worked with a wholesale distributor who sold to cannabis dispensaries.

Zeke, on the other hand, was more inclined to the bigger payoff that comes from increased risk. He had already done a couple of years in prison in the Midwest. That limited his employment options and left him with some connections who were interested in a promising business opportunity. So, instead of keeping his cannabis vaping cartridges in California where they were legal, Zeke would pack some hardside luggage with cases of vape cartridges loaded with liquid cannabis, and fly to a state where cannabis was still illegal and where his products were very much in demand.

I got a call from Zeke late one afternoon, asking if I could take him to SFO. He said his flight was leaving at 10pm, so I agreed to pick him up at seven. Since it would be a night drive, I knew the traffic on I-280 wouldn't be too bad.

I drove up the hill, way up into the woods east of Felton where Zeke and Rocky lived. Once we had Zeke's 2 large suitcases loaded in the back of my Subaru, we were on our way.

Zeke let me know he had checked the price of the ride with Lyft. It came in at $68, so that's how much he was going to pay me. It was clear that he didn't appreciate having a private driver nearly as much as Vinny and Walter did. I wondered if Zeke would even tip me.

Even before we got to Highway 17, Zeke announced he was going to catch some Zs, lowered the passenger seat back, and went to sleep. It was almost an hour later as we were about halfway up I-280 when Zeke awoke with a start, sat up and looked around to get his bearings, then realized he had things he needed to attend to.

Zeke pulled his wallet out of his right rear pocket and began pulling out various things that might identify him in some way. This took a few minutes, as he carefully went through his wallet, pulling out every single card and piece of paper.

When he was done, he showed the stack of cards and papers to me. "You got something you can put these in?" he asked. "Like an envelope or something?"

"Open the glovebox," I said. "There's probably an envelope in there with my registration and insurance. Leave them in the glovebox, and put your stuff in the envelope."

He found the envelope, pulled out the registration and insurance papers, closed the glovebox, then began stuffing his cards and papers into the envelope. Then, as he was about to hand the envelope to me, he said, "Oh shit, wait. I've got to make sure I've got my attorney's phone number."

"Don't you have that on your phone?" I asked.

"I do. But that's my other phone. It's not the one I travel with."

After digging through the papers in the envelope for a couple of minutes, he found the scrap of paper with the phone number. He stuffed it into his shirt pocket, then handed me the envelope again. "Hold on to this stuff for me until I get back."

He said it very nonchalantly, as though I could file his papers and drive him to the airport at the same time. I opened the console between his seat and mine and dropped the envelope in with the packs of facial tissues, aux chords, and my sunglasses.

Not long after, I exited the 280 for our approach to SFO. A few minutes later I pulled up to the curb of the departing passengers loading zone, got Zeke's luggage out of the back and wished him luck. He paid me with 4 twenty dollar bills. I was almost expecting him to ask for change, but he didn't.

Once I got back to Santa Cruz I took the envelope out of the console, sealed it up with some scotch tape, and wrote "Return to Zeke" on the envelope before putting it back in the console.

Zeke's deal must have gone off without too much trouble because he eventually made it back to town. I think he had forgotten about the envelope, as he never called and asked about it. I caught him at home one time when I was picking up Rocky, and I handed Zeke the envelope. He didn't seem to recall giving me his documents for safekeeping.

I took Zeke to SFO another time. He called me at the last minute, and he was running late. He was also pretty drunk.

As we were going over the hill he was on the phone with the airline, trying to reschedule his flight because we were going to be late getting to the airport. I tried suggesting he ask about flights out of San Jose or Oakland on the chance that one of those would work, but he was apparently too drunk to listen to me and talk with the reservations representative at the same time.

In the end, I guess they found a later flight that worked for him. I got him to SFO. He said he'd have to hang out for a few hours, but it would be alright. It was the last time I ever gave Zeke a ride. I was okay with that.

Chapter 17

Weed, Part 3

Will Legalization Kill the Industry?

On a Saturday morning in 2017, as I was taking Walter to the San Jose Airport, I asked him about Proposition 64. That was the initiative, passed by the voters in 2016, that would legalize cultivation, transportation, testing, and retail sales of recreational (as opposed to medical) cannabis products in California beginning in 2018.

I assumed it would give a tremendous boost to the local industry. Walter wasn't so sure. Astute businessman that he is, he had read the fine print in the propositions.

While I wouldn't describe Walter's politics as Libertarian, he had a keen sense that, in the case of the cannabis industry, the more the government got involved, the worse it would be for everybody's business. It turns out he was right, at least at the local level.

First, there were the new excise taxes of 4%. Those taxes were charged at every step of the operation, not just at the final sale. So the grower paid 4%, the trucker paid 4%, the testing lab (where they measure the THC content) paid 4%, the wholesale distributor paid 4%, and the retail store paid 4%. But that was just the excise tax. The retail store also had to pay sales tax.

Yes, all those taxes were eventually passed on to the consumer at the checkout counter. That pushed the retail price up considerably. Meanwhile, the illicit market for weed continued along much as it always had, with no regulation, no testing, and no taxes. As one writer noted, the consumer who knew a guy who knew a guy, really had no incentive to switch to the legal marijuana outlet in 2018, especially when it might cost twice as much as he had been paying for many years.

As a driver, I gave rides to quite a few people who worked in the industry. Outside of Walter's associates, a lot of those rides were to people who worked at dispensaries. For the most part, even as the law was implemented in 2018, they were doing well. Dispensaries were seeing an uptick in business, but not the huge increase predicted by some.

The young men and women who worked the retail counter seemed pretty upbeat. Their old medical cannabis customers were still coming in, and they were seeing some new recreational users, as well.

But every now and then I gave a ride to someone who worked in the back office, and they weren't feeling the euphoria. They could already see the coming impact of the increased taxes and regulation. As prices for the goods they sold continued to rise, it was going to impact their bottom line. Before long, it wouldn't be sustainable.

Some cannabis businesses survived through consolidation or, more often, big businesses buying up struggling, smaller businesses. This allowed them to save on lab costs, paying less for testing, and needing only a single sample test for a large batch of product. A delivery driver is paid the same hourly wage, whether he's delivering in a small van or a big box truck.

There are as many stories as there were people working in the local cannabis industry. As I was finishing up this manuscript, I contacted Vinny, Walter's partner. I described that conversation Walter and I had in 2017, wondering how things were going for them now.

He said Prop 64 had really messed up the economy across the board in Santa Cruz County. Without going into specifics, Vinny said he was "pretty well-connected," so he was surviving. "But it's not like it used to be." As he put it, "It has definitely separated the haves from the have nots."

By well-connected I assume he meant to suppliers, but I didn't push him for more details. Vinnie would occasionally skate into the illegal side of the business, not as a regular thing the way Zeke did, but only when a rare opportunity came along.

Once when I was giving Vinnie a ride home, he asked if I'd be interested in driving a truck to Texas for $30,000. He said he'd give me a referral to a good attorney should I need one. I just had to promise to keep his name out of it. I thanked him, but declined. I had two daughters in high school. They needed to have a father around. Vinnie has a young daughter. He understood.

Back to our discussion about how their business was doing today, Vinnie said, "I'm surprised we've lasted as long as we have. Fortunately, I was able to put some money aside during the good times. Walter, not so much."

It's hard for me to imagine them down on their luck, financially. And if they are, I have no doubt a lot of other folks in Santa Cruz County are, too.

If I seem a bit sentimental about my cash customers who worked in the industry, it's because they were repeat customers who paid me well and, in some cases, became like friends. Years after I left Santa Cruz, Vinnie or Walter would send me a text, just to see how I was doing. I did the same when a big forest fire hit part of the San Lorenzo Valley.

They were the guys who kept me afloat during the slow seasons, and during my last year of rideshare driving, when Uber was squeezing drivers like we were oranges in a juice factory. It's clear I could never go back to being a rideshare driver in Santa Cruz, even if I wanted to. The numbers just wouldn't add up. It's best that I've moved on.

Chapter 18

Trips to the ER, Part 2

Cheaper Than an Ambulance

That late night trip with Sandy was my first trip to the ER as a rideshare driver. It wasn't my last.

I realized there was a good chance it would happen again. But I also knew there was risk involved. Not risk as in legal liability. But I do feel an obligation to do what is right for my passengers. I understand that a ride with me is much cheaper than an ambulance. And sometimes I might even be able to get a passenger to the ER faster than an ambulance can. But I also need to be able to say, "No, you need an ambulance."

I'm not a medical professional. Yes, like all soldiers, I had to successfully qualify in First Aid during basic training, and I've had CPR training, too. But really, that only qualifies me to attempt to keep someone alive until a medical professional can give them the care they need.

With that in mind, I decided I would take a passenger to the ER so long as they were not in cardiac arrest, bleeding out, or just about to give birth. Fortunately, I was able to transport several passengers safely to the ER, and never had to turn anyone down.

There was a guy with an abscessed tooth that was causing him significant discomfort. It was past closing time at his dentist's office on a Friday evening, so he figured his only option was to go to the ER.

There was a guy I picked up at an urgent care facility. They had apparently checked him out, identified the problem, and knew it wasn't something they could adequately deal with there. They told him he should go to the ER.

There was a couple I picked up at the Dream Inn one evening. The woman was having considerable trouble walking, but she managed to get in the car. Earlier that day she had stumbled and originally thought she had just sprained her ankle. But as the pain increased and her mobility decreased, she and her husband began to suspect she may have broken something.

And there was the guy who had tried to take a shortcut to his hotel in Scotts Valley by vaulting over a barbed wire fence. He didn't quite make it, and did some damage to his scrotum. He had managed to stop the bleeding, but felt the need to get it checked out.

Since an ambulance ride to the ER can cost a thousand dollars or more these days, a ride from Uber or Lyft can be a real bargain. Potentially, it also frees up an ambulance for the sort of medical emergency it was intended.

Several months after I took Sandy to the ER, I gave her another ride, this time to a little get-together with some friends. She was doing much better. Thc doctors had finally figured out a successful protocol for her, and it was working very well. I always like to hear a happy ending.

The Calls are Coming from Inside the House

I was driving north on Market Street in the late afternoon. It had been raining on and off for the past couple of days. I was heading home for a short break before starting the night shift. On the side of the road, near the entrance to a condo complex just south of where Market Street goes under Highway 1, was a young woman with a couple of bags. She looked as though she was waiting for a ride, or perhaps a bus, except there is no bus route there. I'll call her Joy.

About thirty minutes later, as I was headed out again, Joy was still standing there. I stopped and asked her if she needed a ride. She said, "I didn't request an Uber."

"I know. But I saw you here a while ago, and you look like you need a ride. It's starting to rain again, so I'm happy to help you out if you need a ride."

"I don't really have any money to pay you with," she said.

"That's okay. You look like you need a ride. It's about to get dark and it's starting to rain. I'm happy to help you out. You don't have to pay me."

She seemed to be assessing her situation, then said, "Can I put my bags in the back seat?"

"Sure. That's fine."

She put her bags in the back seat, then got into the front passenger seat.

"Where do you need to go?"

She found the button to roll down her window. Turning her head, she coughed a bit. It sounded kind of harsh, like a bad cough.

"That doesn't sound too good," I said.

"I know. I think I might have to go to the hospital," she said.

"Is that where you want to go?"

"I'm not sure. I don't know if I should."

That sort of set the tone for our entire conversation. I'm not sure how to describe it, but something about her just felt a little bit off. I've given a lot of rides. People usually know where they want to go.

I headed south on Market, then turned on Water Street, which eventually merges with Soquel near the intersection with Morrisey. She coughed some more. It wasn't raining hard yet, just a slight drizzle. I kept our windows open a bit, hoping some fresh air would minimize the chance of me catching whatever was causing her cough.

She asked me to turn on Morrisey. "My parents live on the other side of the highway."

"Is that where you want to go?" I asked.

"Maybe," she replied. "I'm not sure. My son is there."

She directed me to a residential area between Highway 1 and DeLavega Park. She continued to give me directions. Then I realized she was talking to someone, but it wasn't me. I glanced at her. She was not talking on a cell phone.

She told me to make another turn. It seemed we were just going around a block.

"Were you talking to someone a minute ago?"

"Yes. I was talking to my son. We communicate telepathically, but I have to be close enough. Otherwise it won't work."

"So that's why we're driving around the block?"

"Yes. We passed my parents' house. I was asking him if they were home. He said they are, so I can't go see him right now." She coughed some more.

"Do you want to tell him you might have to go to the hospital?"

"Yes, that's a good idea."

I could hear her talking again.

When it seemed like she was finished with the telepathic conversation with her son I said, "We're close to the hospital. Would you like to go there now?"

"Yes," she said. "I think maybe we should."

A few blocks east of her parents` neighborhood there is a not-quite-secret shortcut to the hospital, a little one-way road barely wider than a bike path that goes through a tunnel and crosses over Arana Gulch before popping out on city streets again. In less than five minutes we were pulling up in front of the door to the ER. I got out and took Joy's bags out of the back seat. She took her bags and walked up the ramp and through the door into the ER.

As I was driving away, I thought about my friend Brad in Flagstaff. He's a social worker at the Flagstaff Medical Center. When there is somebody at the hospital who, after being treated for medical issues, needs to go over to the Behavioral Health unit, Brad is the guy who walks them over there. He has a very calming presence, not the least bit threatening. People tend to go with him willingly. I wondered if there was someone like Brad at the hospital in Santa Cruz.

Chapter 19

Short Takes

Way Out

Unlike most other universities, at UCSC there is no Greek Row, the block or two of fraternity and sorority houses with their identifying Greek letters proudly displayed on the front of each house. Consequently, each of the organizations must find a house somewhere off campus. That's not easy in a town where there are simply not enough bedrooms for everyone who lives here.

A few of the fraternity houses are well-known to me. There are parties there nearly every weekend.

During the weekend at the end of the first week of each quarter, pre-Rush Week begins. This is a week of parties and other get-togethers that allow new students to begin to check out the school's Greek life in preparation for Rush Week. Rush Week is sort of the formal process of students checking out specific fraternities or sororities, and being checked out by those same organizations. At the end of Rush Week, decisions are made about who is invited to pledge which organizations.

That pre-Rush weekend can be a boon for drivers who are up for it. There are parties at the usual locations during the afternoons and again well into the night. The drinking starts early, so most of the students aren't feeling much pain by the time they start asking for an Uber to come get them and take them back home to freshen up and change clothes in preparation for whatever evening festivities are on tap.

I pulled up to one of the fraternity houses late on Saturday afternoon. There were probably 30 or 40 students milling around in front of the house. I noticed a small group of girls not far from where I was parked as I waited for my passengers. Two of them were kissing. It wasn't your average kiss on the cheek between sorority sisters. They were definitely doing some serious kissing, and they seemed to really be enjoying it.

I was distracted for a moment when a guy came up and asked me who I was there to pick up. I gave him the name. He said she was coming.

I looked back at the girls who had been kissing. One of them was walking away from the group. As she left, the girl she'd been kissing said, "You have a really nice ass."

Clearly, fraternity parties are a lot more interesting now than they were when I was in college, at least in Santa Cruz.

Take Care of Your People

On a busy Saturday night, I had just dropped off a passenger in Soquel and was getting back on Highway 1, heading west back to Santa Cruz when I got a request for a pick-up at the 19th Hole. That's a bar at the gated country club neighborhood of Pasatiempo.

The 19th Hole is actually just outside the gate. There are a dozen or so motel rooms there, too. I've picked up numerous passengers there. It's often the site of company holiday parties and the occasional wedding reception.

Pasatiempo is just a mile north of downtown Santa Cruz. I guess the request came to me because I was getting on the freeway. I could get there faster than a driver who was downtown, even though any driver downtown would have been physically closer. Uber's big computer must have these space/time relationships figured out.

I took Highway 1, then the ramp for Highway 17. The Pasatiempo exit is the first exit on 17. I got off at the exit, took the overpass across 17, and pulled into the parking lot of the 19th Hole. I drove right up to the entrance, looked around, and waited, looking for my passenger. Let's call him Jerry.

After a couple of minutes, I pulled out, and drove around the parking lot. No sign of Jerry, so I called him. Jerry said he was just then coming out of the bar. He sounded a little drunk. I drove back to the entrance. Jerry stepped out of the shadows near the entrance and got in the back seat.

Once I verified he was, in fact, Jerry, I touched the Start Trip stripe on the Uber screen on my phone. The destination address showed up on the app. Jerry lived inside the gate.

As we approached the gate, Jerry gave me his code, I punched the buttons, and the gate opened. Even though I was getting good guidance from the Google Maps/Uber/GPS interface, Jerry proceeded to tell me every turn I had to make.

As I pulled up to his house, I could hear what sounded like Jerry pulling some bills out of his pocket.

Before getting out, Jerry handed me a small wad of cash.

"Here," he said. "Take care of your people."

I thanked him. Then he said, "If you ever need anything, call me," and exited the car. I assume he didn't realize I didn't have his phone number. All phone calls and texts between drivers and passengers go through a double-blind system at Uber, so neither knows the other's actual number.

As I pulled away, I reflected on Jerry's rather magnanimous final statements. Once I got back outside the gate and found a safe spot to pull over, I turned on the Subaru's interior light and unwrapped the wad of bills Jerry had given me, half expecting maybe one of them would be a twenty. Instead, Jerry had given me two one dollar bills.

I headed back toward downtown, thinking of all the ways I could take care of my people with a couple of wadded dollar bills.

Dating Advice

One of the interesting things about being a rideshare driver in a college town is having so much contact with young people.

One night I picked up three young women in Scotts Valley. They all sat in the back of my Impreza. They seemed like they were just barely out of high school, so age-wise they might have been 18, maybe 19.

I didn't catch much of their conversation at first. They were talking about boyfriends and dating. I was focused on getting us safely onto Highway 17 southbound, then the transition onto Highway 1 going east. Their destination was a popular Midtown restaurant known as The Crepe Place.

On Highway 1, heading for the Morrissey Boulevard exit that would take us into Midtown, one of them said, "Let's ask our driver."

Another chimed in with, "Yeah Burt, you've probably had a lot of experience. Share some wisdom with us. What have you learned about relationships over the years that you think young women like us should know?"

"You really want to know?" I said.

"Yeah. Lay it on us!"

This was certainly not a conversation I had prepped for. On the other hand, it was an opportunity to perhaps do something good. Maybe I could tell them something their parents and school counselors never had. I was willing to give it a shot.

"Okay ladies, since you asked, here it is. When you're getting to know someone, there are certain conversations you should have before you take off your underwear."

This was greeted with amazement, as though this concept had never once come up before.

"Really? Like, what kind of conversations?"

"Two things. First, before you have sex, you need to discuss who is responsible for birth control, and what that's going to be. And second, again, before you have sex, you each need to discuss if you or your partner have any sexually transmitted infections, you know, STIs."

"Wow. I mean, I'm sure you're right, but like, those aren't always easy things to talk about."

"You're absolutely right. But you know what's even harder to talk about?"

"What?"

"Telling your partner your test came back positive, whether it's for pregnancy or an STI. Because *that* conversation has a bunch of other implications. You'll find that, though it may seem a little awkward at first, having those conversations on the front end is much easier and much less complicated than it is on the back end."

"Oh wow, you're right. Thanks. And thanks for being cool enough to tell us that stuff when nobody else ever has."

About that time, I caught a break in the oncoming traffic, made a U-turn, and pulled up right in front of The Crepe Place.

"Ladies, it has been my pleasure. Have fun tonight, be smart, and don't be afraid to ask the tough questions."

"Thanks, Burt. Hope you have a good night."

Chapter 20

Learning to Say No

I've never been very good at saying "No." Most of the places I worked in the military, I was part of a "Can do" culture. We tended to approach challenging situations by asking, "How can we get this done? How can we make this work? What will we have to do to make this happen?"

I once asked our Air Operations Officer how he made sure we had the aircraft we needed for a particular training operation, since the entire operation depended on having the aircraft that would get us there. He replied, "If we need one bird, I request three. If I ask for three, they'll give us two. If they give us two, one will probably have a problem and won't be able to fly the mission. That leaves us with the one bird we need for the operation." Those were the sorts of creative solutions we thrived on to accomplish our missions.

As a rideshare driver, I tried my best to accommodate my passengers. But there were times when I had to say, "No, I'm not taking this trip."

Sometimes it was not a matter of my personal preferences. For instance, sometimes the requesting passenger will come out with four (or more) friends—and tell me, "It's okay, we can all crunch together in the back seat."

That's when I have to say, "No. Company policy and the law in the state of California say there must be a seatbelt for each passenger and the driver."

I have seatbelts for four passengers—one in the front and three in the back. More than four passengers is a deal breaker. I know I'm going to get responses ranging from hurt feelings to outright hostility. I offer the simple solution. Put in an additional request. Then there will be plenty of seatbelts between the two vehicles. But, of course, that will cost twice as much.

When I first started driving for Uber, there weren't other options. By the time I gave my last ride, many markets offered the Deluxe option, which provided a larger vehicle. With either alternative, an extra five or ten dollars split 5 ways isn't that much. If they all really want to get to where they're planning to go, they'll usually figure it out.

If they can't figure out a way for only four or fewer of them to get in my car, I would just cancel the ride, wish them a pleasant evening, and move on to the next customer.

Then there are the dog owners.

Don't get me wrong. I like dogs. To paraphrase Mark Twain, "If you take a hungry dog, feed him and take care of him, he won't bite you. This is the principal difference between a dog and a man."

Here's what I've learned: Dogs and their owners tend to leave dog hair all over the back seat. Once I got the faux leather seats, this wasn't as much of an issue. But when I still had cloth seats, it was.

And frankly, what I make on an average ride doesn't cover the time and money it takes to thoroughly vacuum the back seat after the ride.

Most passengers who want to bring their dog will call me as soon as I accept the ride to ask if it's okay if they bring their dog with them. I've had passengers tell me the dog would be riding in a travel crate (it didn't). I've had passengers say the dog would stay on their lap the whole time (it didn't).

So after a few experiences like that, when a passenger called and asked if they could bring their dog, I said, "No." I suggested they put in another request. Maybe on their next try they would get a driver who doesn't care. Then I would cancel the trip.

There is one important exception: Service Animals. Per federal law and company policy, a driver can't refuse a passenger who needs to bring their service dog with them.

And frankly, service dogs are the best-behaved dogs I have ever encountered. I keep an old GI laundry bag in the back of my car. It's made of pretty tough cotton material. It covers about half the back seat when I spread it out. The trick is to assess the situation, and get it down on the seat where the dog is going to sit before the passenger and the dog get into the car.

The final category of passengers I'll say "No" to are those who are too drunk to ride in my car. I've only had to do that once.

A little after ten on the Saturday night following my Uber Baptism, a request came in for a pick-up near a popular bar in downtown Santa Cruz. I pulled into a parking space close to the designated pick-up location, near an intersection with crosswalks, but no stop light.

I waited a minute for a passenger to show up. When no passenger arrived, I called the requester. He said they would be right there. He called me a minute later and said they were just about to cross the street.

I looked at the intersection and saw two young men attempting to carry a woman through the intersection in the crosswalk. Once across, I guess they had to stop for a break. They set her down on the sidewalk briefly, then picked her up again and carried her down the sidewalk to my car.

I assessed the situation, remembering the young woman who threw up in my car the previous weekend. One of the guys opened a rear door, and holding her under the arms, started to put her in the back seat. It turned out they weren't even going to ride with her. They were just going to leave her horizontal in the back seat. There were so many reasons why that was a bad idea.

I said, "I'm sorry. I can't transport her in that condition." She couldn't even sit up on her own.

They said, "Please."

I said, "No."

They said, "What are we supposed to do?"

I said, "You wait until she is sober enough to sit up by herself and give coherent directions to her driver. Then you put in another request. I appreciate you trying to help her out, but you've got to do it right. I can't take responsibility for her in that condition."

I canceled the trip and headed for my next request. I wondered if I'd get another request from those guys later that night, but I didn't. I hope she eventually made it home okay.

Chapter 21

Uber Therapy, Part II

Looking for Love in All the Wrong Places

For a while I was driving over the hill several mornings each week. I'd see those surges in Silicon Valley on the app on my phone screen, and convinced myself I could make more money if I was willing to brave the early morning rush hour traffic on Highway 17, just like the hundreds of people who work in Silicon Valley but choose to live in Santa Cruz.

One morning as I was passing Los Gatos (the first town north of the Santa Cruz Mountains), I received a request that took me into a residential area of tract homes on the southside of San Jose.

A woman stepped into my car as soon as I arrived. Her name was Desiree. She appeared to be in her late twenties, maybe early thirties. She had dark skin and black hair. Desiree was one of those unfortunate souls who met a guy at some bar or club, hit it off with him, went home with him, then discovered she had definitely gone home with the wrong guy.

In her case, it turned out the guy was a tweaker, a user of crystal methamphetamine, a "meth head."

I didn't catch the whole blow-by-blow story of her night with the tweaker. All I know for sure is that she wanted to get out of there as fast as she could. I tried my best to oblige.

Desiree was likely a stranger to the south end of the Valley. First she told me to take her to the nearest BART station. BART is the Bay Area Rapid Transit, the light-rail metro train that connects much of the Bay Area. But while BART is great for getting to San Francisco from Oakland and the rest of the East Bay, BART had not yet pushed a line as far south as San Jose.

Once she realized she was nowhere near a BART station, she told me to take her home to where she lived in the East Bay. A few more miles down the highway she changed her mind again and decided she actually wanted to go to her mother's house in Richmond. Richmond is about 50 miles north of San Jose, so we had a lot of time to talk.

In my experience, once a passenger spills out their story, given enough time, they ask me what I think. This is a crucial and sensitive moment for any therapist, Uber or otherwise. It's easy to judge another person based on the mistakes they've made. I try not to do that.

In a branch of the counseling field called Reality Therapy, there are two important questions a therapist asks: (1) What are you doing? And (2) How's that working out for you?

I don't think it's appropriate for a driver to be that direct. People don't pay for a ride so they can be interrogated by a stranger. But I hoped I could nudge her in that direction. To use a couple of film metaphors, it seemed she had gone from an Urban Cowboy scenario to Looking for Mr. Goodbar without realizing it until it was too late.

Perhaps that's the roll of the dice anytime you choose to go home with a stranger. Alcohol plays a part, lowering inhibitions and impairing judgment. Whether it's the longing or the heart or just the plain old libido that gets a person to a singles bar or club, throw down a few drinks and, even though you came in looking for Mr. Right, in many cases you end up leaving with Mr. Right Now.

I said, "I can't make your decisions for you, Desiree. But clearly the position you got into last night didn't feel safe, perhaps wasn't safe at all. You put yourself at considerable risk. You may have to ask yourself how much risk you can accept. Is it worth going through something like you just went through in the hope of hooking up with a guy you're really going to like? And you might want to have that conversation with yourself before you go to the club."

She started to speak, then stopped. I imagined her mental process. *Everybody goes to the club. Everybody wants to meet somebody, either for a hot night or the chance of a long-term relationship. How are you supposed to know if somebody is a tweaker or even some kind of psycho killer when they seem really nice while you're dancing with them?*

"Damn," she finally said. "I mean, that's what clubbing is all about, trying to meet somebody nice, you know, having a good time. But, shit, I definitely don't want to end up in another situation like last night."

"Maybe you could always have a plan B," I suggested. "Always keep your phone handy. If it starts to feel the least bit weird, request an Uber, grab your shoes and coat, and get out the door."

"Yeah, that is probably a good idea. And once I got my phone, I could always dial 911 if, you know, something real bad is going on."

Somebody else might have laughed when they said that. Desiree wasn't laughing. She was dead serious. She was making a plan. I hoped she never had to use it.

Sad Lisa

There are times when it's best for the Uber Therapist to just keep quiet instead of telling the passenger they might want to consider changing the way they go through life. The wise Uber Therapist learns to recognize those times.

I like mornings on Saturdays and Sundays. Everyone's mood seems a bit subdued. That generally goes for mine, too, as I usually drive well past midnight on Friday and Saturday nights.

Most of these weekend morning trips fall into one of four categories. There are the occasional trips to one of the airports. There are people (usually college students) who need a ride to work. There are those who left their vehicle somewhere the previous night because they were smart enough to request an Uber or a Lyft ride home. They need another ride in the morning to get their car or truck. And finally, there are those who went home with somebody, woke up in the morning, realized what they'd done, and are quite eager to get away from there as soon as they can. It's the rideshare equivalent of the Walk of Shame, I guess.

On a Saturday morning I was sitting in a parking lot at Cabrillo College, after I took a guy to work at a restaurant in downtown Soquel. I accepted a request for a pick-up in the Rio Del Mar area.

I found the address, and pulled into the uphill driveway. It took a few minutes, but eventually my passenger came out. Let's call her Lisa.

Lisa was an attractive woman with long blond hair. She wore dark leggings that clung to her shapely legs and bottom, and a tight fitting sheer cream colored top with a neckline that showed off Lisa's significant cleavage. As Lisa got closer, it was clear she was not wearing a bra.

As she was about to get into the car, she wasn't sure if she had her phone. She apologized, and said she'd be right back. Apparently, she hadn't left her phone in the house. She hurried back into the house, discovered it wasn't there, then found it in her purse. Lisa hurried back to the car and got in the back seat.

I could not easily guess Lisa's age. Her Botox lips suggested she could be forty. But her voice sounded more like she was in her twenties.

As I was verifying she was my correct passenger, she started to sound a bit frantic, saying, "Can we please just go now?" Clearly she was anxious to get away from that house as soon as possible. Then she started to cry. "Why am I always so unhappy?" she asked. I didn't get the sense she was actually asking me for an opinion.

I pulled a pack of tissues out of the console and handed them to Lisa. She continued to cry. As we drove the three miles to her house in Capitola, she apologized twice for crying.

"No need to apologize," I said. "Really, it's okay."

Stopping in front of her house, I asked, "Are you going to be okay?"

"Yes," she said. "I really am sorry for crying so much. It's just so hard when it involves heartbreak."

She got out. I watched her enter her home.

I've tried to imagine the series of events that brought Lisa to the point of so much sorrow, and why that sorrow has become such familiar territory. Has she been looking for true love in bars of Aptos and Capitola, a magnet for guys who assume she's good for a one-night stand and aren't really interested in anything beyond that?

That could certainly lead to some perpetual sadness, I suppose. Plenty of room there for the Uber Therapist to tell her all the things she was doing wrong, but it's also good to know when to keep his observations to himself. Just because she's been looking for love in all the wrong places doesn't mean she was looking for advice from her rideshare driver. I hoped she had someone to talk with who would tell her the truth. It couldn't be any more painful than what she'd been doing to herself.

I'll call her Scarlett. She had beautiful red hair.

I picked her up from an upscale Capitola bar. She'd been there celebrating with friends, but they left in another car, heading in a different direction. She looked good in her navy blue evening dress, her red hair down past her shoulders. She was on her cell phone when she got in. She'd had a few drinks. She sat in the passenger seat.

It didn't sound like a very happy conversation. All I really caught was, "Yeah, well thanks a lot," just before she ended the call.

"Damn it!" she said. "That fucking son of a bitch!"

I took the liberty of not interrupting her, starting the drive based on the destination on the app. She was silent for a few moments, then took a deep breath. "I'm very sorry about that."

"No problem," I said. "Just to be sure, could you tell me your name, please?"

"Oh. Of course. My name is Scarlett."

"Thank you, Scarlett. I'm Burt. And we're going to Seaview?"

"Yes, thank you. And again, I am sorry about that little outburst."

"No need to apologize. Is everything okay?"

"Yes and no. I mean, really everything's okay. It's just, it's my birthday, that was my ex calling about some bullshit I don't really need to hear about. And he never once said 'Happy Birthday.' But I should know by now. It's just the way he is."

"I'm sorry to hear that. Happy Birthday."

"Thank you. He's always been like that. It's like he does it just to be mean. And I know I shouldn't let it get to me, but sometimes it still does. When we broke up, he told me he never thought I was very attractive, but he stayed with me because I'm a good fuck. That's the kind of shit he says to me. And I try not to let it bother me, but it still hurts."

"I'm sorry, Scarlett. You deserve better than that."

"Yeah, no shit. He's just jealous because I've got a good job, because I make more money than he does, because I own my own house in Santa Cruz."

"I bet he is. What do you do?"

"I work for a tech start-up in San Jose. I'm one of about a half dozen female vice-presidents in all of Silicon Valley."

I asked her the name of the company. It was one I had never heard of. That's not too surprising. While I've picked-up and dropped-off a lot of folks in Googleland, and the Apple campuses, and to a lesser degree Oracle, Cisco, and Facebook, there are dozens, maybe hundreds, of other tech companies, large and small, that have a presence, if not their corporate headquarters, in Silicon Valley.

You've probably never heard of the company either. But if, as Scarlett said, there are only a half dozen or so corporate vice-presidents in Silicon Valley who are women, I'm going to protect her anonymity by giving the company another name. Let's call it Xanity. Scarlett told me the company has a very tall building in downtown San Jose.

"And I'm guessing," I said, "within an extremely male-dominated industry, you women VPs must all know each other."

"Oh, most definitely. We get together for drinks once a week and talk. I'm mean, really, who the hell else can we talk to?"

"That's good. I'm glad you've got a support system. I imagine there are times when that can be really helpful."

About that time I was turning onto her street. The lighting wasn't great. I slowed a bit. She pointed out the house that was hers. Nice house in a nice neighborhood.

As I pulled into her driveway, Scarlett reached into her purse, found some cash, and handed me a ten dollar bill, a tip for an eight dollar ride.

"Thank you, Burt. It was nice to ride with you." She opened the passenger door and stepped out.

Before she closed the door, I said, "Thank you, Scarlett. I want you to remember something. You're beautiful. Anyone who tells you you're not is an idiot. Happy Birthday." I hope she believed me.

On numerous trips to San Jose over the next few months, I would occasionally look up at the skyline, wondering where that big corporate building might be, the one where Scarlett said she worked. After a while I started to wonder if it even existed, if maybe Scarlett made up all that stuff about being a corporate executive.

Then one weekend I took the train to Los Angeles to visit my daughter at the University of Southern California. The Amtrak doesn't run through Santa Cruz. I rode the bus from Scotts Valley to the Amtrak station in San Jose, then caught the train going south. On the return trip I took the train from LA to San Jose. I was sitting outside the train depot in San Jose waiting for the bus to take me back over the hill. It was early evening. The sky was turning dark.

I happened to glance up toward the San Jose skyline. There in bright lights at the top of a very tall building, I could see the name of the company: XANITY.

Chapter 22

The Kindness of Strangers

The winter of 2016-2017 was epic, truly one for the record books. We got rain. Big rain. Lots of rain. More rain than the Central Coast has seen in years, maybe decades. On the upside, we needed the precipitation. Now the drought is over (for the moment), the reservoirs are full, and the mountains have a good snowpack.

On the downside, floods, mudslides, and fallen trees (really big trees) have taken their toll. Driving any of the northbound roads out of town requires skill and caution. Even more so during a heavy rain. Lots of rain in the mountains makes the ground soft, then causes it to move.

Sometimes the side of a hill collapses, covering two lanes of highway (and any vehicles that happen to be there) with tons of mud. Or, if the slide happens on the side of the highway that is downhill from the pavement, the road just falls away, leaving a large gaping hole where there once was a highway or secondary road.

The rain and wind combined to take down a lot of big trees, both redwoods and eucalyptus. Sometimes when a big tree comes down, it takes a nearby power line down with it, creating an additional hazard.

It was Friday afternoon. I was on the Westside when I got a request for a pick-up in the forested hills northwest of the university. It's a very rural area off of Smith Grade, a 2-lane blacktop road which runs between Empire Grade, west of the university, and the forest community of Bonny Doon.

Rain was falling; not a heavy rain, just enough to remind you it ain't over til it's over. Remote, rural area. Bad weather. It's the sort of request most drivers would turn down. I was feeling heroic, so I took it.

Driving up Empire Grade past the western entrance to the university, I began to question the wisdom of my decision. Pine trees hung over the road on both sides, just waiting for some unsuspecting driver. Maybe I still think I'm a paratrooper. You stand in the door. You go. In this case, you drive on.

Further up the road I found Smith Grade. I made the turn. I looked at my GPS map. There wasn't much but white blank space on either side of the road, occasionally crossed by crooked blue lines indicating streams, streams that were flowing like they hadn't flowed in years.

The pick-up address was on Smith Grade, but the actual GPS location was not. I found a row of mailboxes, one of which had the address on it. There was a side road that headed south into the forest. I followed.

Cell phone coverage is a little like police officers. It's never there when you really need it. So I couldn't call the passenger to ask where the house was. The best I could do was follow roads that looked like they were getting me closer to the pin placement on the GPS map. At best, the GPS was showing me where I was and where it thought the pick-up location was, all against a blank background, with no named roads showing at all.

I was getting close. I had found a couple addresses that were numerically close to the one I was searching for. By then I was off the pavement and on dirt roads. Except they weren't really dirt. They had been mud roads for the past few months. I took another fork which I thought would get me closer. It was muddy and rutted. At one point, my Subaru swerved just a little as the front wheels slipped into a rut in the road, then straightened out. Then the engine died.

I spent about 20 minutes trying to get it started, to no avail. The engine would try to start, would turn over a couple of times, then it would die, sort of like I was out of gas. Except I wasn't. I had more than half a tank. I began to think some muddy water had splashed up onto a sensor, maybe messing with the air/fuel mixture. Then I began to wonder what I was going to do out in the middle of nowhere with a broken-down car in a rainstorm and no signal on my cell phone.

I grabbed my phone, got my rain jacket, locked the car and started walking back the way I had come, hoping to find some kind of help. There was a cabin nearby, but there was nobody there. It looked like it wasn't occupied on a regular basis. Further up the road, I found a larger place. Lights were on, and I could see what looked like a telephone line coming into the house from a nearby wooden utility pole.

I'm not very comfortable just walking up to a stranger's home out in the woods and knocking on the door, but it was getting dark, it was still raining, and it would take me most of the night to walk home. I didn't really see any other options.

A woman answered the door. She was in the middle of a phone call, had a headset on her head and a cast on one foot. She paused her phone conversation to talk with me. I explained my situation. She invited me in.

Once she got off the phone, she introduced herself as Johanna. I told her my story of trying to find a passenger in her little neighborhood tucked deep into the forest, then my car broke down. She was starting to make dinner. She asked if I'd like to eat. I accepted. She asked if I had any dietary restrictions. I told her I used to be a vegetarian, but these days I generally eat whatever I'm served.

She let me use her phone to try to get some roadside assistance. I called both AAA and my insurance company. Surprisingly, my insurance company, USAA, would cover the entire charge for a tow truck. AAA would only pay a portion, costing me over a hundred dollars out of pocket, due to the remote location, several miles from town.

But there was a catch to both. Apparently, tow truck drivers have a rule that their trucks don't leave the pavement. I would have to find a way to get my car out of the mud, and about a quarter mile up the muddy road I drove in on, to get it back on blacktop.

Johanna got some meat, some potatoes, and collard greens cooking, then looked through a directory of the residents of her little forest community. We found the address that corresponded to the one I had been looking for, a nearby neighbor. She called them. Her neighbor said nobody there put in an Uber request. So much for advanced technology.

Johanna and I talked, getting to know one another. She was close to my age. She was divorced. She had the main house (where we were), and 2 other cabins on the property that she rents out. She had retired from the university some years before.

It was getting chilly in her cabin. There was a wood stove in the corner. I offered to get it going. She told me to go for it. She explained she hadn't been able to do it very well since her foot had been in the cast. She showed me where her wood pile was, just outside the back door.

I gathered up some split logs, some kindling, and some old newspaper.

Among some of my camping friends, the art and ethic of building a fire, whether in a wood stove or a campfire, suggests that a well-built fire requires only a single match to get it going. Johanna handed me a box of kitchen matches. I took one match out, closed the box, lit the match, and got the fire going quite nicely. I think Johanna was pleasantly surprised, since I was an acknowledged city dweller (that is, I lived in Santa Cruz, rather than up in the hills).

Not long after that, Johanna served up some food for both of us. It was a very nice gesture, not only inviting a stranger in, out of the rain, but feeding him dinner, too.

After we finished eating, a friend of hers came by. He was delivering a well-made wooden box with a toilet seat attached to the top. I could tell the craftsmanship was excellent (it's actually quite difficult to build a box).

Johanna introduced me to her friend Pablo. The box, she said, was to be the upper portion of a composting toilet.

She explained my predicament to Pablo. He had a truck and said he might be able to help me. So after he and Johanna visited a bit, Pablo and I walked out into the darkness to the front of Johanna's property, and found the road that ran along the western edge of her property, eventually finding my car where I had left it.

It still wouldn't start. Pablo looked at the muddy road and said he didn't think he could get enough traction to pull my car out. But he offered to take me back into town. Before we left, we pushed my car off the road and onto the driveway that led to the vacant cabin so it wouldn't be blocking the road.

Actually, I couldn't imagine anyone being crazy enough to try to drive on that road in those conditions—anyone but me, of course.

We stopped back at Johanna's to let her know the situation. Then we jumped into Pablo's truck, and he took me back to Santa Cruz.

We were sitting at a red light at the intersection of Highway 1 and River Street when a distracted driver hit us from the rear at full speed, knocking Pablo's truck into the intersection. Fortunately there were no cars on River going through the intersection at that moment.

Pablo's truck was totaled, as was the car of the guy who hit us. Somehow none of us were seriously injured. The impact knocked my cell phone out of a cargo pocket of my pants. I didn't realize it until I got out of the truck and headed back to see if the guy who hit us was okay. When I reached for my phone to call 911, it wasn't in my pocket. So I returned to the truck and searched around the floorboard until I found it.

Someone else must have reported the crash, since by the time I found my phone the first responders had begun to arrive. I called my ex. She and one of our daughters came to pick me up.

The next morning, I was getting coffee and a bagel at The Firefly. I was telling the story to Katy, The Firefly's owner, and a couple of other folks who were sitting at the counter. When I explained how I had to find a way to move my car about 400 yards to get it on the pavement before a tow truck would come get it, Katy asked me if her late model Toyota Tundra would do the job. I said I thought it would. She asked Matt, who covers managerial duties while Katy is out of town, if he'd be willing to drive her truck to pull my car. Matt agreed. We planned to meet the following morning.

On Sunday morning I caught a ride to The Firefly, arriving around seven. I'd brought a piece of an old climbing rope (retired some years before). Katy handed Matt the keys to her Tundra, and we were off.

I remembered enough about the road into Johanna's place to locate the rutted road, and we found my car. Matt and I were able to push it from the driveway back to the rutted road and pointed in the right direction, then rolled it back a bit to give Matt room to turn Katy's truck around, so both our vehicles were pointed toward the nearest pavement. We hooked up the rope to the truck and my car. Then with me steering through the ruts, Matt and Katy's truck got us back onto pavement.

We left the Subaru there, drove back to town, stopped at a gas station where I topped off Katy's truck with gas (seemed like the least I could do), and called USAA to send a tow truck. The driver met me at the gas station. I rode with him back to where we'd left my Subaru. He loaded it on his truck, and we took it to the shop where I got my maintenance and repairs done.

I heard from my mechanic on Monday morning. Apparently the swerving from the rut I hit was enough to jar a hose loose. That messed with the engine's oxygen mix. That's why the car died and wouldn't start.

Johanna and I became friends, and I took her to and from the airport numerous times.

Katy eventually sold The Firefly to a local couple and opened another place in South Lake Tahoe. If you're ever there, be sure to check out the Dragonfly Bagel Company.

Chapter 23

The Business End, Part 3

When Your Car is Your Business

My silent partner in all these tales is my car. It's a royal blue 2015 Subaru Impreza. It's the best handling car I've ever owned. It has the tightest turning radius of any car I've ever driven, and that includes the 4 Honda Civics I've owned, and a Subaru Outback. I never gave much thought to a car's turning radius until I became a rideshare driver, frequently executing U-turns to head in the opposite direction to get to my next passenger.

The Impreza was the first new car I purchased in 18 years. My daughters were approaching driving age. I had been driving a 2-wheel drive Toyota RAV-4. We were living in Flagstaff where it snows in the winter. I was making deliveries for DHL in my RAV-4. On a single day in early March, I got stuck on ice 3 times. I knew it was time to consider a different vehicle.

Reviewing the economical 4-wheel (or all-wheel) drive options, I went online to see what Consumer Reports had to say. One thing stood out. The 2015 Subaru Impreza had the best crash test results of all the cars in its class. With my soon-to-be-driving daughters in mind, that played a major role in my decision.

I also did a little research on the best ways to buy a new car. Apparently new car dealerships pay a lot of attention to their End of Quarter sales numbers. My RAV-4 was not in the best shape, being about 15 years old, so I figured I'd have to negotiate its value as a trade-in toward the purchase of a new Subaru.

I took the girls with me to the local dealership so they could give me their input on the comfort and roominess. We test drove a few different Impreza models. There was a blue one I liked, and they concurred.

Then it was time to do the sales talk. I was non-committal about financing, allowing the possibility that I might finance through the dealer, though I knew that was unlikely. We settled on a price for the Impreza. Then it was time to talk about the trade-in. Their used car trade-in guy gave me a number. It wasn't enough. I was polite. I left the salesman my card and said if the manager should change his mind about the trade-in value, please give me a call. That was mid-March.

The salesman called me back on March 29th, 2 days before the end of the quarter. The manager had decided they could pay what I was asking for the trade-in. The next day, I bought the Impreza.

By the end of May that year, we learned that both daughters had gotten low lottery numbers in the annual drawing for the new admissions at Pacific Collegiate School in California. We made the decision to move to Santa Cruz. By July, we had found a 4-bedroom house to rent. Really, it had 3 bedrooms upstairs. I got a room on the ground floor that was probably an office for previous renters. Still, it had a 2-car garage, room for guests, and Branciforte Creek just beyond the back fence.

So at the end of July, with the help of some friends, I loaded up a large rental truck and drove it to Santa Cruz. A friend drove the Impreza there for me, then I drove her back to Arizona.

I had to jump through the hoops of getting a California driver's license and auto registration before Uber and Lyft would approve me to drive for them in California. Once that was accomplished, and I passed my background checks, I could start earning money.

I did pursue another job opportunity. In Flagstaff, I was a subcontractor with DHL, doing what is known as overflow. Apparently DHL has a fairly good idea how much volume their salaried drivers can deliver in a day. Anything over that amount is called overflow.

I had picked up the DHL overflow gig about a year before. Their driver would leave parcels inside my back gate three times a week, and I'd deliver for DHL once I finished my Verde Valley route.

I would file nightly reports online with DHL, both so they would know the status of their parcels, and because that's how I got paid. As a result, I had some contacts in the DHL office in Phoenix who were kind enough to say nice things about me to the DHL office in South San Francisco.

I drove to the South San Francisco DHL office for an interview, and they offered me a job. But I only had to do that drive once to realize I didn't want to commute over an hour each way during Bay Area rush hour traffic in order to get to and from my job.

Ironically, I would do that trip frequently as a rideshare driver, taking passengers from Santa Cruz to the San Francisco Airport (SFO). But I would get paid to do it.

The summer tourist season in Santa Cruz comes to an end around mid-September. When that happened, my weekly earnings took a hit. I started to think about finding a source of additional income.

When I would start driving each morning in Santa Cruz, after dropping my girls off at school, I would see lots of surge pricing in Silicon Valley on the map of the Uber app, indicating drivers were likely making more money there than I was in Santa Cruz.

So I tried heading over the hill a few mornings a week. What I discovered was demand for drivers exceeded supply because drivers were stuck in the rush hour traffic jams, just like everyone else. A single ride might bring in more income than a normal ride in Santa Cruz, but I was giving fewer rides each hour, plus burning time and fuel driving over the hill and back. That was not a solution that would work for me.

Google was making a play in the Bay Area to compete with Amazon, but with a different business model. I applied and, during the winter of 2015-2016, was hired as part of a team of drivers working for a transportation company that got the contract from Google. Mostly we drove box trucks to Costco, loaded them with goods people had ordered on the Google App, and drove them to one of three transportation hubs, where they were off-loaded and sorted for last-mile delivery.

That adventure is a whole other story which I'll save for another time. After a few months I got injured trying to move a loaded pallet without a pallet jack. That got me a workman's comp claim, and nudged me back toward being a full-time rideshare driver again. Besides, the tourist season in Santa Cruz was starting, and the Uber bonuses were looking good. So my Subaru and I were working together again.

Halloween Night of 2016 was a Monday, which made it the last night of the 4-night Halloween weekend in Santa Cruz. On Friday, I noticed what seemed to be a lot of dust in the car. On Saturday at the car wash, I tipped the vacuum guy an extra 10 bucks, asking him to do a really good job on the interior. I guess he did the best he could, but somehow the cloth seats must have absorbed a lot of dust. If I slapped my hand down hard on the seat, more dust came up.

On that Monday night, business was unusually slow for Halloween. But really, the parties had been going on since Friday. It was raining, it was a Monday, and maybe people were just too tired to party anymore. Or maybe they stayed home to watch the Bears play the Vikings on Monday Night Football.

I was parked in my usual spot in front of People's Coffee, with no ride requests and wondering why my car seemed so dusty. I Googled various permutations of "dust in my Subaru." What kept coming up was *Cabin Filter*.

Ironically, when I had the oil changed the previous week, Isaak, my mechanic, said the car was due for a new cabin filter in another 5,000 miles.

I found a video on YouTube on how to change the filter in a Subaru. I drove to the nearest auto parts store. They didn't have the one I needed. The next store did. I pulled out the glove compartment, pulled out the old filter (filthy), popped in the new one, and with only a small amount of effort managed to get the glove compartment to snap back into place.

It took a few hours of running the heater to clear the air, so to speak, but by the next morning the air in the cabin was better.

But I realized that my cloth seats were going to get dusty again, and I wouldn't always be running the heater or the air conditioning with Santa Cruz's relatively mild climate. So I did a little research and found some faux leather seat covers made especially for the 2015 Impreza. I ordered some white ones for front and rear. With a little work I was able to install them myself. And suddenly, I had what looked like white leather seats.

They were nice, easy to clean (in case someone barfed on them), and didn't collect dust the way the cloth seat covers did. That was really all I had hoped for.

But there was an added bonus. I started to get compliments from passengers about what a nice car I had. Just the appearance of leather seats somehow made my Impreza feel somewhat luxurious. I think they cost me around $250. It was a great investment.

A feature of the Santa Cruz area streets are dips that are built where driveways meet streets and where streets meet parking lots. These dips facilitate the flow of water so that, during times of heavy rainfall, the build up of water flowing on the streets is minimized.

One of the reasons that the 2015 Impreza handles so well is because its frame is much closer to the ground than the Outback, Crosstrek, or Forester. However, that low clearance also means the front bumper is likely to scrape the pavement when going through one of those dips unless the driver hits it very slowly and carefully.

I must confess I sometimes did not hit those dips slowly and carefully. Often I was in a hurry to get to my next passengers. I was only getting paid for the time and miles when I had actual passengers in the car, so I would frequently hurry to get to my next pick-up point. As my Mexican taxi driver says, "Tiempo es dinero." Time is money.

So my bumper frequently got dinged. And sometimes the decorative plastic covers that go where fog lights could be installed (but weren't on my car) would get jarred loose and fall on the pavement. I ended up replacing the front bumper twice, and bought several of those fog light covers to replace ones that fell out when I hit a dip too hard.

Because of the mileage I put on the car, I would get an oil change about once a month. Needless to say, I was on a first name basis with Isaak, the owner of the tire and auto repair shop where I got my service done.

I have often joked with my mechanics, both in Santa Cruz and in Arizona, that I was putting their kids through college. During the first half of 2019, it felt like I was putting Isaak's kids through college along with my own.

When I brought the Subaru in for its monthly oil change in mid-January, they noticed the engine was leaking oil from the head gasket and the timing cover. Further inspection showed oil leaking from both sides of the head gasket and there was oil in the intake manifold. Isaak said it would cost less to find another engine with low miles than it would to rebuild the current engine. So that's what we did.

He found a 2016 Subaru engine with only 15,000 miles on it. Its car had been totaled, but apparently the engine was fine. So my Subaru with 170,000 miles on it got an engine with just 15,000. That engine is still running as I write this in 2024. All it cost me was $6,667. That included the oil change.

The car ran fine through the rest of the winter and spring. By mid-May at 186,000 miles, it needed 4 new tires. The tires, wheel balance and alignment, and the monthly oil change ran me $900. Good tires aren't cheap.

A month later at 190,000 miles, the Check Transmission dashboard warning light came on. I needed a new one. Again, Isaak recommended finding a used transmission, which he did, again with relatively low mileage. Total cost for parts and labor: $3,115.

A week later it was time to take the Subaru to San Jose for its annual Uber car safety inspection, during which the lights, signals, wipers, horn, and tires are all checked. But the horn wasn't working, so I took the car to Isaak. I was hoping it was just a fuse or something simple like that. What Isaak discovered was a larger problem, a bad sensor within the steering wheel assembly, which is also where the driver's side airbag is housed. Apparently, that repair was above his pay grade. He took it to the local Subaru dealer. The whole thing cost $890.

Once I got the car back, I drove over the hill to San Jose for the safety inspection. Ironically, it was the only Uber safety inspection I'd been through that I wasn't asked to demonstrate that the horn worked properly.

My last Santa Cruz automotive expense occurred just before my last trip to Arizona. At that point, the Subaru had 196,580 miles on it. I noticed a vibration in the steering wheel that didn't feel like a wheel balance issue. It occurred when I was turning. I didn't want to drive across the desert without having it diagnosed. Isaak checked it out. I needed a new axle assembly. It cost $450. Kyrene and I drove from Santa Cruz to Flagstaff with no problems.

So, excluding regular oil changes or headlight bulb replacements that were not associated with one of those major repairs, my final 8 months of driving in Santa Cruz left me with over $12,000 of credit card debt. I hope Isaak's kids did well in college. I know my girls did.

During my four years and over 10,000 trips as a rideshare driver in Santa Cruz, I put 187,500 miles on my Subaru. That's over 46,000 miles a year.

It now belongs to my daughter Simone who lives on the East Coast. As of this writing in the fall of 2024, it has 215,000 miles on it and is still running.

Chapter 24

The Exit Plan

It happens every year about this time. The weather turns cold (44 in Santa Cruz this morning), and I start thinking about moving south. Way south. South of the Border. Latin America.

Nearly a year before we moved to Santa Cruz, I got this crazy idea that, once the girls were in college, I could head off to somewhere in Latin America, living a comfortable retirement with a low cost of living. I began to research several Latin American countries in both South and Central America.

My online research and conversations with friends, relatives, and random travelers had me looking as far south as Colombia and Ecuador. I considered Panama and Nicaragua. Costa Rica has gotten kind of pricey, sort of the Scottsdale of Central America.

Lately my focus has been Mexico. It's a big country with lots of options, many of them very nice, relatively inexpensive, and as safe or safer than much of los Estados Unidos. Yes, there are drug cartels, but there aren't any random mass shootings at schools, concerts, shopping malls, places of worship or movie theaters. I've kept track of my research on a blog called Somewhere South of Perfect. Nobody reads it but me. That's just fine.

It's not just the weather, of course. I have other motives. With 2 daughters heading off to college soon, I would like to reduce my overhead enough to be able to send money to both of them every month.

They have some college savings. They are both smart and doing well in school. They both could get some need-based and merit-based financial aid. Still, going to college is quite expensive these days.

So, I'll continue my driving gig for a while, at least until I can get the girls settled into college. I'm hoping my cash customers can carry me through the down times, and hopefully offset the many ways Uber keeps trying to squeeze every last nickel out of us drivers.

But it seems quite likely that, before too long, I'll find a way to walk away from all this paradise and these wild and crazy nights, and find a warm and peaceful place South of the Border where maybe I can sit quietly with my laptop and turn these twisted tales into a coherent narrative of my years as a rideshare driver in Santa Cruz.

I wrote another book once, a murder mystery set in Arizona. About ten years after it was published, my former spouse borrowed a copy I'd given to one of my daughters. After reading it she told me, "You know, you're really the only person who could have written that story." I'm guessing this book will be like that, too.

Chapter 25

A Little Night Music

For a few months before moving to Santa Cruz, I worked on Friday nights as a volunteer DJ at a new community radio station in Flagstaff. I loved the opportunity to craft song sets together, drawing upon several decades and multiple styles and genres of music.

I grew up in Phoenix. Not long after I graduated high school, a couple of free-form music stations emerged there, allowing their DJs to play whatever they wanted, and the DJs did. For a few years in the early 1970s, the Valley of the Sun had one of the most knowledgeable and sophisticated radio audiences in America. I was part of that audience.

In addition to the Beatles, Rolling Stones, and Crosby, Stills, and Nash everyone else was hearing on the radio, we got to hear early Bonnie Raitt, John Prine, Jimi Hendrix, T-Rex, Pink Floyd, Fairport Convention, Richie Havens, Loudon Wainwright III, Mott the Hoople, Traffic, Fraser & DeBolt, John Hartford, Joni Mitchell, Blind Faith, and Quicksilver Messenger Service, to name just a few.

Like all good things, those days didn't last forever. It didn't take long for free-form radio to morph into corporate-owned album-oriented radio, that eventually became the corporate playlist radio, where DJs were only needed to cue up the next pre-recorded commercials.

During my years in the army, I learned it was possible to hear some new music by tuning into college radio stations. That's where I got introduced to artists such as Nanci Griffith, Steve Earle, Shawn Colvin, Sarah McLachlan, Robert Earl Keen, Keb Mo, Dwight Yokum, and the Indigo Girls.

In Santa Cruz there is a very special radio station. KPIG is an independently-owned community station that is quite quirky, free form, and includes weekly programs with both local musicians and those on tour performing live in the studio. KPIG became another source of new music for me.

While music streaming had been a thing for a while, first on Pandora, then Spotify, I found neither of them sufficiently intuitive for my somewhat eclectic tastes. By the time I started as a rideshare driver, I knew I wanted control over the music.

I would listen to KPIG online at home, and I would make notes about new artists and songs that I liked. Then, just like when I was a DJ, I would create song sets I'd want to hear while driving, keeping in mind what my passengers might like as well as what wouldn't be appropriate for a captive audience.

At first I would transfer the music files from my computer hard drive onto my Android phone, which meant while I was driving my phone was handling the music, the navigation apps (Maps or Waze), the Uber or Lyft apps, and any phone or text communication. It's a testament to modern technology that a single phone could do all that.

When I upgraded to an iPhone, for a while I was using two phones, with music still pumping through the Android, and the iPhone running the other apps.

Eventually the Android died. Unable to move the music files to the iPhone, I opted for streaming, but with a catch. I already had access to a lot of music I had purchased through Amazon. So I started using that music to create playlists on Amazon Music. And that worked. I could just choose which of my playlists I wanted to hear and press PLAY.

The first of those Amazon playlists I created, simply titled Nights #1, started out with CS&N singing David Crosby's *Long Time Gone.* I tend to like live performances, so next up was Shawn Colvin's solo acoustic cover of Tom Waits' *The Heart of Saturday Night.* That flowed into three KPIG favorites: Sun Volt's *Joe Citizen*, Hayes Carll singing *Times Like These*, and Nicki Blume's *To Rise You Gotta Fall.*

I can recall reflections of a full moon bouncing along the tops of the waves while listening to Susan Tedeschi singing *Midnight in Harlem* as I drove along East Cliff Drive. That led into Lake Street Dive doing *You Go Down Smooth*, followed by The Little Willies doing a beautiful cover of Dolly Parton's country classic *Joleen.*

I'd let Lucinda Williams raise the temperature a few degrees with *Righteously*, then follow it with Roy Schneider and Kim Mayfield's lovely ballad *Angels Along the Road.* I'd blend in some blues with Keb Mo' singing *I Remember You* and Bonnie Raitt doing *Love Sneaking Up on You.*

That seemed the perfect spot for Jackson Browne's *That Girl Could Sing.* Then came Santana's *Smooth* (with Rob Thomas from Matchbox Twenty doing the vocals), followed by Carrie Rodriquez (on vocals and fiddle) performing her autobiographical *Z,* and Richard Thompson doing a live version of *1952 Vincent Black Lightning.*

Dave Alvin & the Guilty Women added a little Cajun flavor with *Marie, Marie.* James McMurtry kept things light with *Just Us Kids.* Then, from the Bob Dylan - 30th Anniversary Concert, Roger McGuinn, Tom Petty, Eric Clapton, George Harrison, Neil Young, and Dylan himself all sang Dylan's *My Back Pages.*

By then, we were pretty deep into the night, so it was time to wrap things up with the Grateful Dead doing *Don't Ease Me In,* Kenny Wayne Shepherd singing *We All Alright,* and Will Ray's tribute to his favorite town, *Santa Cruzin'* (spoiler alert: "There were hippies everywhere, and reefer in the air"). The Indigo Girls singing *Galileo* seemed a perfect way to end.

To be honest, the song lists were all made up of music I wanted to hear. For me, they were a key ingredient in getting me through hours of sitting in my car, driving from pick-up to drop-off, over and over, while doing my best to give my passengers good, safe, friendly service.

For the most part, my passengers enjoyed the music. I remember specific compliments for songs by Sarah McLachlan, Sade, and Steely Dan.

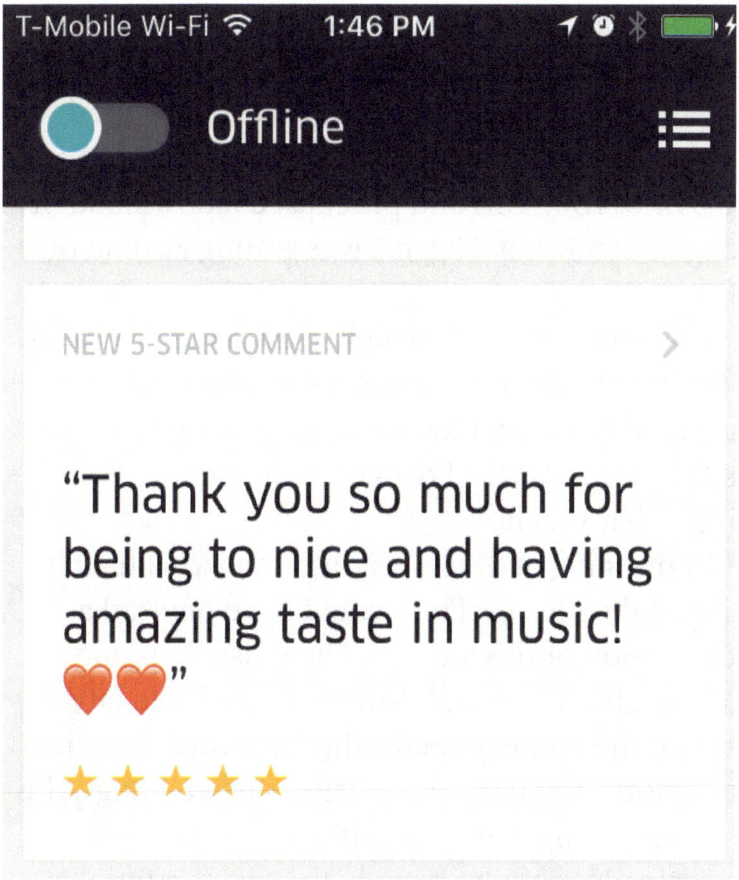

One young woman literally wanted to hear something she could dance to in the car. I handed her my Android so she could pick her own music. After a few seconds of scrolling through the list of artists and songs, she said, "Oh my god! You actually got both versions of "*Walk This Way,*" meaning both the original Aerosmith recording, and the cover by Run-

DMC. She picked the Aerosmith version. She danced as well as she could with her seatbelt on.

I once gave a ride to Laurie Roberts, who was the program manager and daytime 10 to 2 DJ at KPIG for several years. She was surprised when I played a song she'd never heard before—*The Wrong Girl* by an Australian artist named Missy Higgins. I explained how I used to check out indy musicians on Myspace. I liked *The Wrong Girl* and purchased a download of the song. Soon Missy Higgins was getting airtime on KPIG.

One passenger, let's call her Caroline, had once been married to the road manager of a famous LA musician. She and I ended up singing along to Little Feat's *Dixie Chicken* as I drove her home one evening. That was fun!

In finishing up this memoir, I realized I hadn't mentioned the music. That would have omitted a vital dimension of this story. The music kept me going through the ups and downs, through the city streets and the country roads, the highways, the trips to the airports, the bars, the parties, the weddings, the beaches, the campus, the hospital and, eventually, home again. Hopefully, it made the nights a little sweeter for my passengers, too.

Chapter 26

The Business End, Part IV

Like a Rat on a Wheel

In her revealing August 2018 piece in *Vanity Fair*, Susan Fowler, who shook the tech industry with a blog post about the climate of sexual harassment she experienced at Uber, shared a stunning anecdote about the corporate mindset that drives how the company manipulates drivers to do more for as little as possible.

"A few weeks into my tenure at Uber, where I started as a software developer just a year after graduating from college, still blindly convinced I could make the world a better place, a co-worker sat down next to my desk. "There's something you need to know," she said in a low voice, "and I don't want you to forget it. When you're writing code, you need to think of the drivers. Never forget that these are real people who have no benefits, who have to live in this city, who depend on us to write responsible code. Remember that." I didn't understand what she meant until several weeks later, when I overheard two other engineers in the cafeteria discussing driver bonuses—specifically, ways to manipulate bonuses so that drivers could be "tricked" into working longer hours. Laughing, they compared the drivers to animals: "You need to dangle the carrot right in front of their face." Shortly thereafter, a wave of price cuts hit drivers in the Bay Area. When I talked to the drivers, they described how Uber kept fares in a perfectly engineered sweet spot: just high enough for them to justify driving, but just low enough that not much more than their gas and maintenance expenses were covered."

That certainly confirmed some things I had suspected about Uber's operations for a while. Keep in mind Uber has never made a profit—*EVER!* This may come as a shock to readers who don't follow the industry, but Uber burns through hundreds of millions of venture capital dollars every single quarter. [2]

For a few years, Uber founder Travis Kalanick was able to keep the cash flowing in by telling investors that an Uber fleet of self-driving cars would soon be on the road, drivers could be eliminated altogether, overhead would go down, and profits would magically appear. And that worked for a while.

But the reality finally sank in that driverless cars (or autonomous vehicles, as the industry likes to call them), might happen someday, but they were definitely not "right around the corner." And, of course, drivers will tell you that a driverless car won't know when a passenger has puked all over the back seat. Imagine being the next passenger to slide into that seat.

The reality is the business model doesn't really work. Everybody points to Amazon as the success story of a business that lost money for a few years before it became insanely profitable. It's a nice example of a company for which increased volume of sales, and with it, the economy of scale, finally paid off. That business model worked.

But that model doesn't really apply to Uber. While Uber has some constant costs in the form of administration (including a whole bunch of computer engineers and programmers, we must assume) that won't grow as the number of riders increases, the bulk of operational overhead is paid out to drivers in the form of signup bonuses, payments for each trip, and

---

[2] Written in 2019. Uber first reported an annual profit in 2023.

incentives to drive more, especially at certain times in certain areas.

With early investors getting itchy to cash out, especially as Uber prepared for the initial public offering (IPO) of its stock, the company was anxious to show some sort of glide slope in operating costs they could point to as a pathway to profitability.

There are really only two pathways to such profitability. One is to charge passengers more—not likely while locked in war with Lyft for passenger loyalty. The other is to pay drivers less.

But both companies are also in a war over driver loyalty. They know many drivers do both services. So how does Uber get drivers to stay on its app and off the Lyft app? The answer is two bonus programs: The Consecutive Trips bonus and the Quest bonus.

The Consecutive Trips bonus pays the driver a set amount (usually between $9 and $12) for doing 3 Uber trips in a row in certain locations during certain times of the day. The catch to consecutive trips is the driver must stay on the Uber app over the course of those 3 consecutive trips, including the time between trips. That means the driver can't be on the Lyft app at the same time.

Now $12 isn't much of an incentive, but it is technically possible to drive 3 sets of 3 consecutive trips over the course of a busy 2-hour period, during either morning or afternoon rush hour. That's $36 extra for a 2-hour window, which can turn a $64 morning into a $100 morning. That's how Uber manipulated its drivers to stay on the Uber app and off the Lyft app.

By the way, I don't mean to paint Uber as the bad guy and Lyft as the good guy here. While Lyft might have been cooler in the beginning of this rivalry, and didn't have the abrasive Mr. Kalanick as its CEO, the economic realities of the rideshare market have caused the two companies to grow closer and closer together, at least in terms of the way they manipulate their drivers.

The Quest Bonus is a bonus for giving a certain number of rides during a given number of days. I first started seeing it in the summer of 2016 when it was called the *Power Driver Plus.* At the time it was for an entire week, Monday through Sunday. There were several levels of total rides, with increasingly higher bonus amounts for higher numbers of rides. Of course, pushing hard to rack up a lot of rides with Uber tended to keep me off the Lyft app. But at that time, Lyft's incentives were based on both time of day and location (and all those locations were over the hill in Silicon Valley and the San Francisco Peninsula). Driving primarily in the Santa Cruz area, there wasn't much incentive for me to drive for Lyft when I was making good bonus money with Uber.

The weekly *Power Drive Plus* bonuses in the Bay Area, which included Santa Cruz, looked like this:

> 30 rides per week, $50 bonus -- adds $1.67 to each ride.
> 50 rides per week, $100 bonus -- adds $2.00 to each ride.
> 80 rides per week, $250 bonus -- adds $3.13 to each ride.
> 100 rides per week, $375 bonus -- adds $3.75 to each ride.
> 120 rides per week, $500 bonus -- adds $4.16 to each ride.

Those numbers remained constant for about 4 months. I liked knowing that if I did a certain amount of work each week, I could count on a certain reward for my hard work and my loyalty to the company. The bonus for giving 80 rides could turn a $750 week into a $1,000 week. And if something came up (as things often do when you're a parent), there's still a little bonus money to be made in the next level down.

Generally, I was averaging about $6 per ride after Uber took its cut. You can see how adding $2, $3, or $4 to the amount per ride could make a significant difference in my paycheck each week. I really liked that. During that summer and early fall I was doing well financially.

That fall, I learned from another driver that his *Power Driver Plus* program split the week into 2 segments, Monday through Thursday and Friday through Sunday. That seemed strange to me, but I was driving around 40 hours each week to hit those big bonus numbers. I can see how having 2 different bonus periods each week could be advantageous to a part-time driver. But nobody ever asked me if I might prefer one over the other. Uber didn't ask us "partners" for our preferences.

That information also made me aware that Uber was individualizing its weekly bonus offerings. In essence, Uber was beginning to test each driver's responses to various incentives.

Then, around the end of October, Uber changed the program. The announcement came in an email with a subject line that said, *"Introducing a new weekly offer just for you!"* (Yes, it actually ended with an exclamation point.) Apparently paying out those hefty bonuses was taking its toll on Uber's bottom line. Or perhaps we were just moving into the next phase of Uber's rat lab experiment. Having conditioned us drivers to respond to the bonus stimuli, Uber began to change the numbers each week.

I can only imagine that Uber's engineers were implementing a new behavior modification algorithm. They were testing whether a driver who would give 100 rides for an extra $375 each week would do the same for $325, and, if so, would a driver also do it for $250. They were looking for that "sweet spot" mentioned by Susan Fowler above, seeking the minimum possible reward to keep each driver running on that rat lab wheel, thus keeping the gears of the big Uber machine turning every single day, but costing the company less to keep them turning.

On October 30, 2016, the bonus promotion changed from 5 levels to 3. The news arrived in an email with a subject line of: *"Your special weekly offer!"* (Yes, with the exclamation point again.)

30 trips per week, $80 bonus.
70 trips per week, $250 bonus.
120 trips per week, $500 bonus.

I completed 77 trips and got the $250 bonus.

Despite fewer levels, the numbers actually looked good. But why only 3 levels? What was up?

Then here was my November 6 special weekly offer:

50 trips per week, $100 bonus.
90 trips per week, $300 bonus.
120 trips per week, $500 bonus.

I completed 90 trips and got the $300 bonus.

Still not bad, but again, why the change in levels (again!)? And 90 trips earns $300, when 100 used to earn $375. I might have done 10 more trips for an extra $7.50 per trip.

Over the next two months, payouts at all three levels trended lower, with the number of rides required at each level frequently changing. This often made week-to-week comparisons difficult, but the trend was unmistakable.

The January 15, 2017, bonus offer continued the trend of lower payouts and the glide slope of the top end bonus.

10 trips per week, $40.
40 trips per week, $60.
120 trips per week, $430.

The January 29 bonus offer continued to mess with the goal numbers.

30 trips per week, $60.
70 trips per week, $120.
120 trips per week, $430.

I should also point out that, during this period, Uber was getting an unusual amount of bad press (during a time when Uber was generally getting a lot of well-deserved bad press anyway), much of it focused on Uber founder and CEO Travis Kalanick being part of Trump's Business Advisory Council. That ignited the movement to "Delete Uber," urging people who have used Uber to dump the app from their phones. This was also when Trump was implementing the travel ban on Muslims, refugees, and other foreigners, many of whom were students.

Lyft responded by pledging to donate a million dollars over 4 years to the ACLU to fight the travel ban. I was still driving for both Lyft and Uber at the time, and for a few months there was a definite uptick in Lyft rides over Uber.

Another factor in the number of rides given was the weather. That winter the Central Coast of California got hit with wind and rain that went on for weeks. The ground became saturated, the wind blew down hundreds of large redwood and eucalyptus trees (many taking power lines down with them), there was a huge and deadly mudslide blocking much of Highway 17 for weeks, and nearly all the secondary roads in Santa Cruz County sustained significant damage from whole sections of road sliding down into a gulch or creek bed.

Many people who normally commuted over the hill to work at tech companies in Silicon Valley made arrangements to telecommute, only going to the office when necessary. While the rideshare business didn't grind to a halt, the weather contributed to a slower than usual slow season.

It was during February of 2017 that my car broke down on a muddy road off Smith Grade northwest of Santa Cruz. I wrote about that in Chapter 21, *The Kindness of Strangers.* With my car out of action for a few days, I wasn't driving at all.

On Groundhog Day Uber changed the rat lab rules again. My week got split into two separate bonus periods. There was Monday through Thursday, and then there was the weekend. Uber introduced it in an email on February 2, 2017:

> *Coming next week: More opportunities to earn, more chances to qualify. Starting Sunday, your offer emails will have a new look and feel. But don't worry, they'll still include the Boost fares and personalized offers (Quest) you know and love.*
>
> *You'll receive an email including weekday offers for trips between 4am Monday and 4am Friday, and another on Thursday, including weekend offers for trips between 4am Friday and 4am Monday.*

This was the split week version of the bonus I'd heard about from another driver a few months before. Now Uber was testing to see how far I would chase the carrot.

Here's how the first one looked:

Monday - Thursday

35 trips, $105.
75 trips, $250.

This initial offering looked pretty sweet. Then again, they always do.

But I gave very few Uber rides that week, as most of the requests came through the Lyft app.

A lot of the folks in Santa Cruz, especially the college students, took the *Delete Uber* movement seriously. So with fewer requests coming in on the Uber app, I was spending more time than usual driving for Lyft.

My Lyft numbers surged beginning in January, 2017, and made up a significant part of my income through September, when I made almost as much with Lyft as I did with Uber. But beginning in October, the *Delete Uber* movement must have lost steam. My Lyft income took a significant drop, from $849 for September, to $283 in October, to $173 in November.

My bank records indicate I didn't drive for Lyft after November, 2017. Uber's manipulations to keep me on the Uber app and off the Lyft app must have worked. A comparison of my weekly direct deposits for 2017 and 2018 indicates no decrease in my total average weekly earnings by driving only for Uber.

With both my daughters scheduled to graduate high school in May of 2018, several weeks were not big money-makers for me, as we visited universities in southern California as well as out of state. Both girls had been accepted to multiple schools. The trips gave them a chance to visit the campuses before making their individual decisions about where they would ultimately go to college.

Just like the previous Uber Quest Bonus, the split week payout numbers started trending lower as Uber continued its drive toward profitability. I don't want to overload the reader with pages of payout data. I'll add it to the appendix at the back of this book for anyone who is really interested.

The BOHICA Trifecta

Toward the end of 2018 and rolling into 2019, Uber implemented 3 changes that deserve special mention. Not only did each one tend to reduce what a driver took home at the end of the week, but Uber tried to make each new change sound like they were doing drivers a favor.

With these 3 changes coming in rapid succession, the term BOHICA certainly seems apropos. I first heard "BOHICA" in the Army during the 1980s while attending the Pathfinder course at Fort Benning, Georgia. It joins that select group of acronyms that describe the experience of being part of a large, impersonal organization, where policies that may look just peachy from the top, look very different from the bottom. Think FUBAR or SNAFU. If you're not familiar with BOHICA, pronounced: bo HEE ka, it's an acronym for Bend over – here it comes again.

Change 1: The New Surge [Dollar amounts instead of multiples]. Here's what the email said on December 4, 2018.

> *No more guesswork*
>
> *We're moving from fare multipliers to dollar amounts so your surge earnings are more clear from the start.*

Uber said it was doing this to make it easier for drivers to see how much they would make.

Does Uber assume its drivers are stupid and can't do arithmetic? Could Uber have missed the fact that the Uber app runs on smartphones, and every smartphone has a calculator that can easily do multiplication? Or better yet, "Hey Siri, how much is seven times three point five?"

Previously a surge was expressed as a multiple of the normal fare amount. A surge of 4.5 would mean the cost for the ride would be four and a half times the normal fare. A driver might get 75 percent of the increased fare, with Uber keeping the rest. So if a normal fare is $10, the surge fare would be $45. That could mean an extra $25 for the driver.

With the new surge expressed as a dollar amount instead of a multiple of the regular fare, Uber would be able to show a $10 surge on the app, expecting drivers would assume that $10 is MORE than the normal rate multiplied times 4.5. *BOHICA!*

Change 2: Time vs Mileage [Paying more for time, less for miles]. Historically, the cost of a ride, a portion of which is paid to the driver, is based on Uber's computation of the distance of the requested trip and the amount of time the trip should take. The new change pays drivers more for time and less for mileage.

The problem with paying more for time and less for miles is **wear and tear on a vehicle is the**

product of mileage, not time. Thanks, Uber. *BOHICA!*

Change 3: The Gamble

Like a Rat on a Roulette Wheel! Pick a color, pick a number. Maybe you'll win, maybe you'll lose.

In late September of 2018, I got an email with a topic line that said:

Maximize Your Weekend Earnings

The more you drive, the more you earn. Check the Promotions tab to track your progress on your 7-day Quest. You'll earn the highest amount in your Quest selection.

You decide how much to drive and how much to earn within your chosen Quest.

This would appear to be a very positive response to driver frustration over the constantly changing Quest bonus numbers. Except, this was from Uber.

Surreptitiously buried in the email was the catch. Yes, you get to choose the Quest goal and cash reward that suits your goals and driving schedule for the week. But if you don't reach your chosen Quest goal, you don't get the next lower reward. What you get is zero. Nothing. Zip. Zilch. Nada! *BOHICA!*

There are many factors that influence how many trips a driver can do during a given period of time. Most of these are outside the driver's control. Longer distance trips lead to fewer rides. Weather can change passenger plans and create unexpected traffic jams.

Even so, the driver must place a bet. Bet on the lower level, you'll get something, but not as much. Bet on the higher level, hoping to get more, and you might get nothing at all. Every time a driver gets nothing, Uber gets closer to profitability. Not exactly a win-win. If Uber were honest with its "partners," it should have called this plan, "*Choose Wrong, Lose Long, Partner.*"

For me, and I assume for most drivers, the bottom line for the driver got significantly worse as Uber tried to make the company's bottom line attractive to investors. For drivers, that's money that pays rent, puts food on the table, and covers the necessities of raising a family, in addition to keeping the car running well. It only stretches so far.

I'm not a psychologist, so I won't generalize about other drivers. But I found this "all or nothing" gambit increased the stress of being a rideshare driver. I have no doubt that additional stress increases the likelihood I would make a mistake at some point. When mistakes are made, bad things can happen on the road. Bad things on the road can impact cars, drivers, and passengers.

Fortunately, I had a network of cash customers who kept me afloat financially, as driving for Uber got progressively worse. A review of my calendar for

2018 and 2019 shows that over time, it was filling up with rides booked in advance by my private clients.

Then, in the summer of 2019, my life took a change that got me off the wheel of the Uber rat lab.

Chapter 27

Truth or Dare, Part 2

At that point in my rideshare career, in March of 2017, between Uber and Lyft, I'd given about 2,500 rides. Nobody had ever said, "Let's play Truth or Dare" before. Nobody. But hey, it was Saturday night. She seemed nice. I was game.

"Okay," I said. "You start."

"Truth or Dare?" she asked.

"Truth."

"Have you ever shot your pants?"

I wasn't sure what she meant. "I think you need to define that for me."

"Have you ever soiled your boxer shorts—I'm sure you wear boxers, not briefs—not just a skid mark, but enough that it could actually fall down your leg?"

She was wrong. I wear briefs. It's funny because she sounded so sure of herself. I didn't correct her. *But really, we're gonna share potty secrets? WTF?*

"Yes," I said. Then I told her the story of driving through Oak Creek Canyon (south of Flagstaff) on a delivery route on a cold morning. I was in serious need of a bathroom break. I knew there were public restrooms at the Indian Gardens Visitors Center, if I could just hold on until then. I almost made it, but not quite. Everything let loose as I rushed into the restroom, before I could get to the toilet. So I cleaned up as well as I could, left the soiled undies in the trash, and bought some new underwear at the JC Penny's in Cottonwood.

My turn. "Truth or Dare?"

"Truth."

"How old are you?" It's a delicate question with some women. Not with Cassandra, apparently.

"That's too easy. I'm 40. In fact, I just turned 40 last week."

"Well, congratulations and Happy Birthday!"

"Thank you. Truth or Dare?"

"Truth."

"Okay," she began her new question, "if you knew someone was doing really evil, horrible things, would you murder them?"

"If I had the opportunity to intervene, I'd do whatever had to be done."

"Sure, but suppose it isn't just a one-time thing, but has been going on for a while. Could you murder the perpetrator?"

"You mean like the German Army officers who detonated a bomb in an attempt to kill Hitler? Yes, I think I could do that. Though in truth, I try not to do things like that these days. Still, these are pretty strange times."

"Yes," she said. "They are. That's why I asked. I've been wondering the same thing about myself."

We could have started talking politics, but we didn't. We had a game to play.

"Your turn," she said.

"Truth or dare?"

"Dare."

I couldn't really think of anything else. "Okay. I dare you to kiss me at the end of the ride."

She paused for a moment, then said, "Why does it always end up going in that direction?"

Interesting that she didn't say, "Why does it always have to get sexual?"

"Hey, I don't know. I've never played this before?"

"Really?"

"Yes. I always thought it was the province of teenage girls at sleepovers."

"But it doesn't have to, does it?"

I didn't reply, having no real answers at that point.

"I wonder," she said, "if it's a result of it just being you and me, here in the front seats, playing Truth or Dare. There's nobody else to witness what happens and make a big deal of it."

She paused for a moment, thinking about it. She laughed.

"What's so funny?"

"I was just remembering something," she said. "Years ago I was teaching in Bogota. I was young, 23 or maybe 24. I thought I would teach them to play Truth or Dare. Big mistake. These were teenagers, with all the associated hormones."

"And you were hot."

"Yes," she agreed without hesitation, "I was hot."

"So, perhaps it's just inherent in the game, or perhaps in our natures?"

"Perhaps," she said.

About that time we were approaching her destination. It was a facility where they cared for her grandfather. She came to Santa Cruz periodically to visit.

I turned into the driveway that ran along the side of the building. I stopped the car. It was dark. She surprised me, turning toward me, putting her arms around my neck and shoulders, she pulled me toward her and kissed me. It was an intimate kiss, and she was not in a hurry. To steal a line from Prince, *She was a girl who knew how to give a kiss.*

When the kiss was finally over, I said, "Thank you. You are a very good kisser."

She picked up her bag and got out of the car, leaving me with a wonderful memory and a few lines in my head that would become a song.

I gave another ride, a fairly long one that ate up the better part of an hour, then returned to the corner of 17th and Brommer, parking in front of People's Coffee, like I often do. I'd been there less than five minutes when I got another request for a pick-up. It was from Cassandra.

So I took her to the Metro Center where she could catch her bus back to San Francisco. Along the way she was full of ideas of using various sensory triggers (sight, sound, and smell) for the next time she would visit. Her grandfather was having trouble remembering things about his life and his family. She was hoping those sensory triggers would help him remember. I hope she was right.

There was no mention of the kiss. When she got out at the Metro Center to catch her bus, she thanked me and, in parting, said, "Have a good life." It's an expression I'd heard years before. It has an implied finality to it, acknowledging the likelihood you will never see one another again.

I will hold on to the memory of your sweet kiss
Like it was my favorite souvenir.

From the song *Favorite Souvenir*

Epilogue

By the beginning of the summer of 2019, my daughter Simone had completed her first year at USC. Her sister Kyrene, who graduated high school in 2018, had turned down partial scholarships from Chapman, Loyola Marymount, and the University of Arizona, choosing to stay an additional year in Santa Cruz, taking classes at Cabrillo College while she considered her options.

I had moved into a cottage in Brookdale, a community in the San Lorenzo Valley between Ben Lomand and Boulder Creek, which put me closer to Walter's home and business, as well as many of his business associates.

Kyrene had just been accepted to the animation program at San Jose State for the Fall 2019 semester, but decided she didn't want to (as she put it) "spend her days at a computer terminal writing code to make a dog's leg move." Instead, she told her mother and me that she wanted to return to Flagstaff, where she grew up and still had close friends, to attend Northern Arizona University (NAU), where she had also been accepted.

I still owned a condo in Flagstaff. Serendipitously, the tenant who had been renting it gave notice that he would be moving out at the end of June.

The proximity of those two decisions was too strong to ignore. I realized if Kyrene and I moved back into my condo, then, as my dependent, she should be eligible for in-state tuition at NAU. That would reduce the amount of financial aid we would have to swing.

So we made a plan. I gave notice to my landlady in Brookdale. Kyrene broke things off with the guy she had been dating. I rented a big truck large enough to hold all her stuff and mine. We loaded it up and drove it across the desert and into the mountains of northern Arizona, the reverse of the trip we had made four years before.

Kyrene's mom drove my Subaru to Arizona a few days later, and we accompanied Kyrene to orientation for new students and their parents. I had been a new student at NAU several decades before, and it had changed quite a bit (more students, lots of new facilities, co-ed dorms, pronouns).

Kyrene's mom flew back to Santa Cruz and found a little place to live that she just loved in the redwood forest outside of Scotts Valley. She continues to work remotely as a health coach for the faculty and staff of a Bay Area university. When USC shut down its campus due to COVID-19, Simone moved in with her mom and attended classes remotely until the campus opened again in 2021.

Just a few weeks after returning to Flagstaff (pre-COVID), I accepted a delivery gig from a former client. Part of it involved picking up lab samples in the Verde Valley and bringing them to a lab in Flagstaff.

Not long after that, the COVID-19 pandemic hit. Much of the other delivery work fell off significantly, but the number of lab samples I'd pick up and deliver began to explode as soon as Covid tests became available.

While many people were either working remotely from home, or had no work to go to—schools, restaurants, movie theaters, and other retail establishments, as well as whole sectors of the travel and tourist industry were shutting down—I spent my days going in and out of hospitals, clinics, and small collection labs, picking up their samples and bringing them to the big lab for analysis. Personal Protective Equipment (PPE) became a line item on my budget.

I kept that up for over a year. And by 2021, I was ready to retire. Both daughters were doing well in college. Simone was scheduled to graduate in May of 2022; Kyrene the following year (like her father, she changed majors a few times).

I let my client know I planned to stop driving for him at the end of May 2021. That sort of worked, except I was still in town, so he would call and ask me to fill in for a few days or train a new driver. I finally had to cut that cord at the end of the year. I knew if I was outside the United States, he'd realize I was no longer available.

I officially retired on New Year's Eve 2021. On New Year's Day, I flew to San Diego, crossed the border into Tijuana, Mexico, and caught a flight to Mazatlán.

It is now 2025, and I'm wrapping up this little manuscript from my apartment in Oaxaca, Mexico, where I've been living on and off for most of the past 3 years.

> *Though I've been up and down this coast*
> *I still love you the most*
> *Goodnight, Santa Cruz*
>
> From the song *Goodnight, Santa Cruz*

Appendix

For those who really care about the numbers, I have included more details about my experience with Uber's Quest Bonus program and the BOHICA Trifecta in this appendix to create a more complete narrative. Some of the information here was also included in Chapter 26. I apologize for any redundancies, but I hope it provides more context to the numbers.

The "Gig Economy" is still in its formative years, with the parameters still being worked out in city councils, courts and state legislatures across the country, as well as the supply and demand realities on the streets. Maybe a social scientist will find this information helpful someday.

The Quest Bonus is a bonus for giving a certain number of rides during a given number of days. I first started seeing it in the summer of 2016 when it was called the *Power Driver Plus.* At the time it was for an entire week, Monday through Sunday. There were several levels of total rides, with increasingly higher bonus amounts for higher numbers of rides. Of course, pushing hard to rack up a lot of rides with Uber tended to keep me off the Lyft app. But at that time, Lyft's incentives were based on both time of day and location (and all those locations were over the hill in Silicon Valley and the San Francisco peninsula). Driving primarily in the Santa Cruz area, there wasn't much incentive for me to drive for Lyft when I was making good bonus money with Uber.

The weekly *Power Drive Plus* bonuses in the Bay Area, which includes Santa Cruz, looked like this:

30 rides per week, $50 bonus -- adds $1.67 to each ride
50 rides per week, $100 bonus -- adds $2.00 to each ride
80 rides per week, $250 bonus -- adds $3.13 to each ride
100 rides per week, $375 bonus -- adds $3.75 to each ride
120 rides per week, $500 bonus -- adds $4.16 to each ride

Those numbers remained constant for about 4 months. I liked knowing that if I did a certain amount of work each week, I could count on a certain reward for my hard work and my loyalty to the company. The bonus for giving 80 rides could turn a $750 week into a $1,000 week. And if something came up (as things often do when you're a parent), there's still a little bonus money to be made in the next level down.

Generally, I was averaging about $6 per ride after Uber took its cut. You can see how adding $2, $3, or $4 to the amount per ride could make a significant difference in my paycheck each week. I really liked that. During that summer and early fall I was doing well financially.

That fall, I learned from another driver that his *Power Driver Plus* program split the week into 2 segments, Monday through Thursday and Friday through Sunday. That seemed strange to me, but I was driving around 40 hours each week to hit those big bonus numbers. I can see how having 2 different bonus periods each week could be advantageous to a part-time driver. But nobody ever asked me if I might prefer one over the other. Uber didn't ask us "partners" for our preferences.

That information also made me aware that Uber was individualizing its weekly bonus offerings. In essence, Uber was beginning to test each driver's responses to various incentives.

Then, around the end of October, Uber changed the program. The announcement came in an email with a subject line that said, *"Introducing a new weekly offer just for you!"* (Yes, it actually ended with an exclamation point.) Apparently paying out those hefty bonuses was taking its toll on Uber's bottom line. Or perhaps we were just moving into the next phase of Uber's rat lab experiment. Having conditioned us drivers to respond to the bonus stimuli, Uber began to change the numbers each week.

I can only imagine that Uber's engineers were implementing a new behavior modification algorithm. They were testing whether a driver who would give 100 rides for an extra $375 each week, would do the same for $325, and, if so, would a driver also do it for $250. They were looking for that "sweet spot" mentioned by Susan Fowler in her *Vanity Fair* article, seeking the minimum possible reward to keep each driver running on that rat lab wheel, thus keeping the gears of the big Uber machine turning every single day, but costing the company less to keep them turning.

On October 30, 2016, the bonus promotion changed from 5 levels to 3. The news arrived in an email with a subject line of: *"Your special weekly offer!"* (Yes, with the exclamation point again.)

30 trips per week, $80 bonus
70 trips per week, $250 bonus
120 trips per week, $500 bonus

I completed 77 trips and got the $250 bonus.

Despite fewer levels, the numbers actually looked good. But why only 3 levels? What was up?

Then here was my November 6 special weekly offer:

50 trips per week, $100 bonus
90 trips per week, $300 bonus
120 trips per week, $500 bonus

I completed 90 trips and got the $300 bonus.

Still not bad, but again, why the change in levels (again!)? And 90 trips earns $300, when 100 used to earn $375. I might have done 10 more trips for an extra $7.50 per trip.

Here was my special weekly offer for November 13:

30 trips per week, $60 (down from $80 just 2 weeks before)
70 trips per week, $200 (not bad--80 trips used to earn $250)

120 trips per week, $500 (the top end is holding steady)

Again, I completed 77 trips, but this time I only got $200.

Here's how my special offer looked on November 20:

30 trips per week, $60
70 trips per week, $150 (a noticeable drop from the previous week)
120 trips per week, $500

Got my 70 trips, and got only $150, a significant drop from the $250 just a few weeks before.

On my November 27 special offer, it looked like at least the number of trips was beginning to stabilize.

30 trips per week, $60
70 trips per week, $180
120 trips per week, $500

Only did 40 trips. Perhaps I was losing my motivation.

Here's the offer from December 4:

30 trips per week, $60
70 trips per week, $150
120 trips per week, $500

Only did 42 trips. Maybe I was trying to send Uber a message.

Here's how it looked on December 11 (no change from the previous week):

30 trips per week, $60
70 trips per week, $150
120 trips per week, $500

I stepped up and did 70 trips. Got the $150 bonus again.

On December 18, there was a significant change from previous offers. The 70 trips level hit a new low, and, for the first time, the 120 trip bonus went down. Merry Christmas, Drivers!

30 trips per week, $60
70 trips per week, $120
120 trips per week, $450

I went out of state that weekend to spend the holidays with some friends and family.

Here was the special offer for the week of December 26 through New Year's Day:

35 trips per week, $90
85 trips per week, $250

In fairness, New Year's Eve is one of the 2 biggest nights of the year for most drivers (the other being Halloween—at least in Santa Cruz). So most drivers were getting plenty of money from a high volume of trips, plus lots of surge money, too. Uber probably knew drivers already had sufficient motivation.

I returned to Santa Cruz in time to drive on Friday and Saturday (New Year's Eve). In those two nights, I gave a total of 40 trips and made $435, over $300 of that on New Year's Eve. In fact, on just 3 of those trips, Uber collected $178 just in surge charges (about 75% of that was passed along to me). One of those was a 7.8 surge, the highest I had ever seen on my app. Uber collected an $81 surge charge on that one 12-minute, 6.5 mile trip, on top of the normal charge of $12.

Rolling into 2017, perhaps in response to my low ride numbers for the previous 2 weeks, the special offer numbers dropped, too.

20 trips per week, $40
50 trips per week, $110
120 trips per week, $470

That week I completed 51 trips for that whopping $110 bonus.

The January 8th special offer seemed to continue the trend of lowered bonus payouts.

10 trips per week, $40
40 trips per week, $90
120 trips per week, $470

I must have been hungry, completing 64 trips that week, getting the $90 Quest bonus, for a weekly total of $530.

The January 15 bonus offer continued the trend and the glide slope of the top end bonus.

10 trips per week, $40
40 trips per week, $60
120 trips per week, $430

With diminishing incentive, I responded accordingly, doing just 43 trips.

The January 22 bonus offer moved a little, sort of like a counter offer.

20 trips per week, $40 (apparently Uber knew I was going to drive more than 10 trips)
50 trips per week, $110 (Uber ups the offer to see if I'll drive 50)
120 trips per week, $430

Late winter and early spring are the slow times for drivers in Santa Cruz, and my numbers certainly reflected that. I only drove 43 trips that week, snagging that big $40 bonus.

The January 29 bonus offer continued to mess with the goal numbers.

30 trips per week, $60
70 trips per week, $120

120 trips per week, $430

I'm not sure why, but I didn't work Monday through Thursday that week, and my numbers for Friday, Saturday, and Sunday were mediocre, at best. Maybe I was sick for a few days. I just can't recall. At any rate, I only drove 30 trips, but at least that got me the $60 bonus.

I should also point out that during this period, Uber was getting an unusual amount of bad press (during a time when Uber was generally getting a lot of well-deserved bad press anyway), much of it focused on Uber founder and CEO Travis Kalanick being part of Trump's Business Advisory Council. That ignited the movement to "Delete Uber," urging people who have used Uber to dump the app from their phones. This was also when Trump was implementing the travel ban on Muslims, refugees, and other foreigners, many of whom were students.

Lyft responded by pledging to donate a million dollars over 4 years to the ACLU to fight the travel ban. I was still driving for both Lyft and Uber at the time, and for a few weeks there was a definite uptick in Lyft rides over Uber.

Another factor in the number of rides given was the weather. That winter the Central Coast of California was hit with wind and rain that went on for weeks.

Many people who normally commuted over the hill to work at tech companies in Silicon Valley made arrangements to telecommute, only going to the office when absolutely necessary. While the rideshare business didn't grind to a halt, the weather contributed to a slower than usual slow season.

It was about this time when Uber changed the rat lab rules again. My week was now split into two separate bonus periods. There was Monday through Thursday, and then there was the weekend. Here's how it was introduced in an email on February 2, 2017:

> *Coming next week: More opportunities to earn, more chances to qualify. Starting Sunday, your offer emails will have a new look and feel. But don't worry, they'll still include the Boost fares and personalized offers (Quest) you know and love.*
>
> *You'll receive an email including weekday offers for trips between 4am Monday and 4am Friday, and another on Thursday, including weekend offers for trips between 4am Friday and 4am Monday.*

Here's how the first one looked:

> Monday - Thursday
>
> 35 trips, $105
> 75 trips, $250

This initial offering looked pretty sweet. Then again, they always do.

But I must have had something else going on. I only drove 2 Uber trips, and those were on Thursday. No bonus for me.

A lot of the folks in Santa Cruz, especially the college students, took the *Delete Uber* movement seriously. So with fewer requests coming in on the Uber app, I was spending more time than usual driving for Lyft.

The Gamble

Like a Rat on a Roulette Wheel! Pick a color, pick a number. Maybe you'll win, maybe you'll lose.

In late September of 2018, I got an email with a topic line that said:

> *Maximize Your Weekend Earnings*
>
> *The more you drive, the more you earn. Check the Promotions tab to track your progress on your 7-day Quest. You'll earn the highest amount in your Quest selection.*
>
> *You decide how much to drive and how much to earn within your chosen Quest.*

This would seemingly be a very positive response to driver frustration over the constantly changing Quest bonus numbers. Except, this was from Uber.

Surreptitiously buried in the email was the catch. Yes, you get to choose the Quest goal and cash reward that suits your goals and driving schedule. But if you don't reach your chosen Quest goal, you don't get the next lower reward. What you get is zero. Nothing. Zip. Zilch. Nada! *BOHICA!*

Here's how a typical weekday (Monday through Thursday) Quest Bonus looked in March of 2019:

Option 1: 100 - 110 Trips

100 Trips, $65
110 Trips, $110

Option 2: 90 - 100 Trips

90 Trips, $55
100 Trips, $100

Option 3: 80 - 100 Trips

80 Trips, $40
100 Trips, $65

Option 4: 70 - 90 Trips

70 Trips, $35
90 Trips, $55

Option 5: 60 - 80 Trips

60 Trips, $30
80 Trips, $40

Option 6: 50 - 60 Trips

50 Trips, $25
60 Trips, $30

Option 7: 40 - 50 Trips

40 Trips, $20
50 Trips, $25

Option 8: 20 - 30 Trips

20 Trips, $10
30 Trips, $15

Here's how a typical weekend Quest Bonus looked in March of 2019:

Option 1: 80 - 90 Trips

80 Trips, $125
90 Trips, $185

Option 2: 70 - 90 Trips

70 Trips, $90
90 Trips, $185

Option 3: 60 - 80 Trips

60 Trips, $60
80 Trips, $125

Option 4: 50 - 60 Trips

50 Trips, $25
60 Trips, $60

Option 5: 40 - 50 Trips

40 Trips, $20
50 Trips, $25

Option 6: 30 - 40 Trips

30 Trips, $15
40 Trips, $20

Option 7: 20 - 30 Trips

20 Trips, $10
30 Trips, $15

Notice that if, for the Friday through Sunday period, a Driver chooses Option 2, and drives 80 Trips, the payment is $90 (the amount of the 70 trip bonus), not the $125 bonus for 80 Trips had Option 1 been the choice.

And, if the Driver chooses Option 2, and for some reason isn't able to do 70 trips, the driver gets no bonus at all.

If Uber were honest with its "partners," it should have called this plan, "*Choose Wrong, Lose Long, Partner.*"

ABOUT THE AUTHOR

Burt Hays is one of the pen names used by the author, who currently resides in the state of Oaxaca in Mexico. He writes songs, fiction, and tales that are mostly true.

He was a rideshare driver in Santa Cruz and the San Francisco Bay area from 2015 through 2019, and during that time he really did give over 10,000 rides.

Mr. Hays has a bachelor's degree in Sociology from Northern Arizona University, and attended graduate school at Georgia State University while stationed in the Atlanta area with the US Army.

He previously published the murder mystery *Asphalt Desert Blues* under a different name, and has been a contributor to *Rock & Ice* magazine. He avoids corporate social media these days, but can be found on Bluesky. @bluesman66.bsky.social

Made in United States
Troutdale, OR
11/03/2026

56844377R00156